A Basic Mathematics Primer

Written to cater for practice in Arithmetic, Algebra and Geometry in the early stages of these subjects.

The book contains a wealth of examples and can usefully be used at different levels throughout the school where reinforcement of basic skills is required.

A
Basic
Mathematics
Primer

Elizabeth Miller, M.A.
Penilee Secondary School, Glasgow.

**HAMILTON
PUBLISHING**

First Published 1982 by
The Hamilton Publishing Company Limited
12 Colvilles Place
Kelvin
East Kilbride
Glasgow

I.S.B.N. 0 946164 00 2

Reprinted 1983

Illustrations by Craig Peacock

Produced in Great Britain by
Thomson Litho East Kilbride Scotland

List of Contents   Page

Arithmetic

Addition and Subtraction

We are going to add and subtract whole numbers in the exercises in this section. Be careful to set your numbers into their correct columns.

Exercise 1

Find answers to:

1. 324+17	2. 74+106	3. 9210+1	4. 23+794
5. 27+270	6. 321+101	7. 84+846	8. 234+20
9. 16+176	10. 8416+1279	11. 23+19	12. 421+826
13. 804+729	14. 73+7032	15. 123+1016	16. 21+179
17. 32+3026	18. 43+3792	19. 17+1072	20. 31+3012
21. 31+106	22. 281+627	23. 16+160	24. 2+24
25. 33+13	26. 1001+201	27. 396+284	28. 29+29
29. 161+293	30. 321+6241	31. 154+42	32. 68+69
33. 125+201	34. 91+92	35. 27+39	36. 82+29
37. 43+44	38. 91+106	39. 2084+16	40. 123+3812

Exercise 2

Find answers to:

1. 31+274+38	2. 62+61+69	3. 3+201+26
4. 3+13+23	5. 34+86+1921	6. 9+99+999
7. 346+2+29	8. 32+64+128	9. 3846+2+21
10. 23+91+998	11. 928+246+32	12. 23+203+692
13. 2+202+2002	14. 84+864+924	15. 0+16+329
16. 31+41+51	17. 38+318+612	18. 26+84+21+9
19. 214+204+23	20. 32+312+204+6189	21. 3+18+6172
22. 824+61+3	23. 18+19+607	24. 13+29+264
25. 38+308+69	26. 1+18+27	27. 33+0+319
28. 24+92+106	29. 112+234+16	30. 39+93+106
31. 84+26+29	32. 302+103+17	33. 89+94+106

34.	$24+204+27$	35.	$16+38+102$	36.	$18+28+92$
37.	$20+40+400$	38.	$106+21+29$	39.	$304+1+3$
40.	$28+27+24$				

Exercise 3

Now for **subtraction**. Remember to 'pay back' after you have 'borrowed'.
Find answers to:

1.	$729-3$	2.	$106-34$	3.	$49-21$
4.	$3268-106$	5.	$236-128$	6.	$2184-1769$
7.	$892-624$	8.	$19-0$	9.	$3218-2914$
10.	$1762-999$	11.	$324-9$	12.	$526-84$
13.	$880-627$	14.	$10\,106-9214$	15.	$3216-123$
16.	$4846-1$	17.	$3291-0$	18.	$724-62$
19.	$18-17$	20.	$32\,167-31\,864$	21.	$18-2$
22.	$301-16$	23.	$34-17$	24.	$94-72$
25.	$301-100$	26.	$27-19$	27.	$38-24$
28.	$102-84$	29.	$3-2$	30.	$1-0$
31.	$17-14$	32.	$826-261$	33.	$325-198$
34.	$192-94$	35.	$789-586$	36.	$1001-94$
37.	$1003-1001$	38.	$28-27$	39.	$2861-1964$
40.	$201-199$				

Exercise 4

Subtraction again. You must decide here, which number is the **bigger** one. Take the smaller number from the bigger one.

1.	$231\,;46$	2.	$584\,;52$	3.	$59\,;284$	4.	$391\,;306$	5.	$9821\,;234$
6.	$42\,;474$	7.	$56\,;506$	8.	$39\,;309$	9.	$0\,;23$	10.	$268\,;286$
11.	$38\,;94$	12.	$5\,;5002$	13.	$826\,;894$	14.	$32\,;306$	15.	$12\,;1869$
16.	$94\,;9142$	17.	$392\,;324$	18.	$32\,;0$	19.	$389\,;388$	20.	$274\,;277$
21.	$100\,;99$	22.	$32\,;94$	23.	$106\,;102$	24.	$100\,;1$	25.	$32\,;19$
26.	$3204\,;111$	27.	$39\,;84$	28.	$102\,;98$	29.	$341\,;308$	30.	$64\,;63$
31.	$94\,;97$	32.	$77\,;73$	33.	$846\,;121$	34.	$101\,;94$	35.	$7\,;10$
36.	$74\,;103$	37.	$181\,;193$	38.	$69\,;69$	39.	$123\,;321$	40.	$1067\,;14$

Multiplication and Division

In this section we will practise multiplication and division of whole numbers.

Exercise 5

Work answers to:

1. 23×4	2. 382×8	3. 28×21	4. 276×11
5. 302×6	6. 826×3	7. 84×16	8. 241×9
9. 314×7	10. 82×12	11. 208×5	12. 69×3
13. 624×6	14. 3296×4	15. 81×11	16. 2840×5
17. 222×12	18. 3216×3	19. 86×4	20. 248×12
21. 18×6	22. 302×4	23. 19×3	24. 102×5
25. 18×2	26. 94×8	27. 67×4	28. 39×3
29. 642×8	30. 381×5	31. 123×4	32. 624×8
33. 17×4	34. 191×3	35. 18×3	36. 101×11
37. 174×10	38. 33×9	39. 1004×2	40. 1235×5

Exercise 6

Worked examples: (a) $18 \times 2 \times 3 = 36 \times 3 = 108$

(b) $2 \times 9 \times 4 = 18 \times 4 = 72$

(c) $5 \times 4 \times 8 = 20 \times 8 = 160$

Work answers for:

1. $18 \times 4 \times 2$	2. $3 \times 6 \times 24$	3. $3 \times 4 \times 8$	4. $2 \times 6 \times 7$
5. $3 \times 18 \times 20$	6. $14 \times 11 \times 2$	7. $3 \times 88 \times 2$	8. $16 \times 2 \times 8$
9. $7 \times 14 \times 6$	10. $3 \times 21 \times 2$	11. $84 \times 2 \times 5$	12. $16 \times 2 \times 5$
13. $33 \times 2 \times 4$	14. $61 \times 3 \times 5$	15. $44 \times 2 \times 8$	16. $16 \times 1 \times 1$
17. $3 \times 2 \times 14$	18. $601 \times 2 \times 1$	19. $308 \times 3 \times 2$	20. $16 \times 4 \times 7$
21. $8 \times 2 \times 9$	22. $6 \times 8 \times 4$	23. $8 \times 9 \times 3$	24. $2 \times 17 \times 2$
25. $3 \times 8 \times 8$	26. $5 \times 5 \times 2$	27. $3 \times 5 \times 1$	28. $5 \times 8 \times 4$
29. $3 \times 16 \times 2$	30. $4 \times 17 \times 3$	31. $3 \times 8 \times 10$	32. $8 \times 19 \times 4$
33. $5 \times 2 \times 2$	34. $6 \times 6 \times 4$	35. $6 \times 2 \times 6$	36. $6 \times 8 \times 3$
37. $6 \times 9 \times 2$	38. $8 \times 4 \times 2$	39. $16 \times 8 \times 8$	40. $4 \times 8 \times 12$

Exercise 7

Work answers to:

1. 23×41	2. 32×24	3. 50×69	4. 28×11	5. 23×26
6. 29×34	7. 83×99	8. 21×22	9. 32×18	10. 21×21
11. 32×32	12. 29×31	13. 28×18	14. 28×44	15. 26×16
16. 29×82	17. 23×11	18. 28×25	19. 23×21	20. 27×24
21. 18×18	22. 19×18	23. 18×16	24. 16×16	25. 11×11
26. 8×19	27. 19×21	28. 32×33	29. 16×94	30. 82×14
31. 18×21	32. 19×27	33. 64×18	34. 62×24	35. 29×31
36. 32×27	37. 31×18	38. 29×41	39. 16×12	40. 14×14

Exercise 8

Work answers for the following divisions: show the remainder **clearly**.

1. $23\div4$	2. $762\div2$	3. $281\div5$	4. $692\div8$	5. $2164\div11$
6. $94\div9$	7. $236\div9$	8. $2914\div12$	9. $1268\div3$	10. $17\div2$
11. $141\div1$	12. $18\div7$	13. $391\div8$	14. $64\div12$	15. $268\div7$
16. $214\div4$	17. $82\div6$	18. $831\div6$	19. $864\div9$	20. $724\div7$
21. $27\div5$	22. $18\div5$	23. $161\div2$	24. $39\div12$	25. $18\div9$
26. $11\div2$	27. $19\div4$	28. $301\div5$	29. $69\div7$	30. $28\div11$
31. $39\div8$	32. $61\div4$	33. $65\div7$	34. $68\div9$	35. $29\div1$
36. $39\div4$	37. $22\div11$	38. $144\div11$	39. $144\div12$	40. $104\div5$

Exercise 9

Sometimes we write, say, $\dfrac{17}{4}$ which means $17\div4$.

So $\dfrac{17}{4}=4+$remainder 1; $\quad\dfrac{29}{2}=14+$remainder 1.

$\quad\quad\;\downarrow$ Quotient $\quad\quad\quad\quad\downarrow$ Quotient

Give quotients and remainders for:

1. $\dfrac{18}{4}$	2. $\dfrac{94}{6}$	3. $\dfrac{72}{9}$	4. $\dfrac{83}{4}$	5. $\dfrac{106}{9}$	6. $\dfrac{24}{9}$	7. $\dfrac{92}{8}$
8. $\dfrac{111}{12}$	9. $\dfrac{33}{8}$	10. $\dfrac{41}{5}$	11. $\dfrac{303}{7}$	12. $\dfrac{62}{7}$	13. $\dfrac{369}{8}$	14. $\dfrac{39}{12}$

15. $\dfrac{102}{11}$ **16.** $\dfrac{110}{9}$ **17.** $\dfrac{1}{6}$ **18.** $\dfrac{4}{3}$ **19.** $\dfrac{7}{5}$ **20.** $\dfrac{132}{7}$

Exercise 10

Still on division. Copy and complete the following.

1. $5\overline{)31}$ ___ +

2. $8\overline{)234}$ ___ +

3. $3\overline{)64}$ ___ +

4. $7\overline{)51}$ ___ +

5. $11\overline{)122}$ ___ +

6. $2\overline{)13}$ ___ +

7. $12\overline{)29}$ ___ +

8. $12\overline{)100}$ ___ +

9. $6\overline{)87}$ ___ +

10. $11\overline{)183}$ ___ +

11. $3\overline{)91}$ ___ +

12. $12\overline{)600}$ ___ +

13. $8\overline{)541}$ ___ +

14. $11\overline{)77}$ ___ +

15. $9\overline{)320}$ ___ +

16. $2\overline{)84}$ ___ +

17. $8\overline{)421}$ ___ +

18. $9\overline{)1000}$ ___ +

19. $11\overline{)72}$ ___ +

20. $9\overline{)406}$ ___ +

Exercise 11

We have **certain rules** when we have a **mixture** of signs: for example $2\times4+5$ has a multiplication sign **and** an addition sign.

We **multiply before** we add.

We also **multiply before** we subtract.

So $2\times4+5$ becomes $8+5$ which equals 13, and $2\times4-5$ becomes $8-5$ which equals 3.

Work answers to:

1. $3\times5+1$ **2.** $5\times5+1$ **3.** $5+3\times2$ **4.** $7+6\times5$

5. $3\times3+2$ **6.** $5\times7+1$ **7.** $3\times9+5$ **8.** $5+9\times11$

9. $2\times8+1$ **10.** $8\times3+2$ **11.** $1+9\times7$ **12.** $2+8\times6$

13. $5+3\times4$ **14.** $5\times8+9$ **15.** $7+7\times7$ **16.** $11+10\times1$

17. $8+6\times8$ **18.** $2+3\times2$ **19.** $7+7\times9$ **20.** $12+12\times12$

Exercise 12

Work answers to:

1. $3\times8-1$ **2.** $4\times6-3$ **3.** $5\times9-8$ **4.** $2\times8-2$

5. $3\times9-5$ **6.** $8\times11-6$ **7.** $19\times2-17$ **8.** $81\times4-10$

9. $33 \times 3 - 1$ **10.** $29 \times 4 - 6$ **11.** $81 \times 9 - 17$ **12.** $31 \times 5 - 101$

13. $84 \times 6 - 321$ **14.** $2 \times 2 - 0$ **15.** $16 \times 10 - 94$ **16.** $18 \times 9 - 71$

17. $99 \times 9 - 161$ **18.** $34 \times 3 - 81$ **19.** $201 \times 7 - 64$ **20.** $82 \times 4 - 150$

Exercise 13

When we have a mixture of signs, for example **division** and addition, we **divide before** we add. We also **divide before** we subtract.

Work answers to:

1. $5 + 16 \div 2$ **2.** $18 \div 3 + 7$ **3.** $8 + 99 \div 9$ **4.** $99 \div 3 + 1$

5. $12 \div 2 + 19$ **6.** $81 \div 3 + 18$ **7.** $7 \div 1 + 3$ **8.** $21 \div 7 + 19$

9. $84 \div 12 + 9$ **10.** $100 \div 5 + 47$ **11.** $93 \div 3 + 102$ **12.** $72 \div 12 + 8$

13. $8 \div 4 + 1$ **14.** $6 + 12 \div 6$ **15.** $12 + 100 \div 10$ **16.** $18 \div 6 + 34$

17. $27 \div 9 + 18$ **18.** $144 \div 12 + 8$ **19.** $132 \div 11 + 1$ **20.** $164 \div 8 + 13$

Exercise 14

Work answers to:

1. $56 - 8 \div 2$ **2.** $100 - 30 \div 6$ **3.** $8 - 12 \div 6$ **4.** $24 - 12 \div 4$

5. $39 - 84 \div 7$ **6.** $100 \div 10 - 3$ **7.** $210 \div 3 - 35$ **8.** $88 \div 11 - 6$

9. $93 \div 3 - 24$ **10.** $17 - 14 \div 2$ **11.** $82 \div 2 - 39$ **12.** $34 \div 2 - 17$

13. $5 \div 5 - 1$ **14.** $1 - 16 \div 16$ **15.** $602 - 38 \div 2$ **16.** $25 - 10 \div 2$

17. $34 - 20 \div 5$ **18.** $82 - 81 \div 9$ **19.** $21 - 3 \div 3$ **20.** $20 \div 10 - 1$

Exercise 15

Write the following fractions in words.

1. $\frac{1}{2}$ **2.** $\frac{1}{4}$ **3.** $\frac{3}{4}$ **4.** $\frac{1}{8}$ **5.** $\frac{3}{8}$ **6.** $\frac{5}{8}$ **7.** $\frac{7}{8}$

8. $\frac{1}{16}$ **9.** $\frac{3}{16}$ **10.** $\frac{5}{16}$ **11.** $\frac{7}{16}$ **12.** $\frac{9}{16}$ **13.** $\frac{11}{16}$ **14.** $\frac{13}{16}$

15. $\frac{15}{16}$ **16.** $\frac{1}{3}$ **17.** $\frac{2}{3}$ **18.** $\frac{1}{5}$ **19.** $\frac{2}{5}$ **20.** $\frac{3}{5}$

Exercise 16

How many **halves** are there in:

1. 1   2. 2   3. 3   4. 4   5. 5   6. $1\frac{1}{2}$ 7. $2\frac{1}{2}$

8. $3\frac{1}{2}$ 9. $4\frac{1}{2}$ 10. $5\frac{1}{2}$ 11. $6\frac{1}{2}$ 12. $7\frac{1}{2}$ 13. $8\frac{1}{2}$ 14. $9\frac{1}{2}$

15. $10\frac{1}{2}$ 16. 6   17. 7   18. 8   19. 9   20. 10

Exercise 17

How many **quarters** are there in:

1. 1   2. 2   3. 3   4. 4   5. 5   6. $1\frac{1}{4}$ 7. $2\frac{1}{4}$

8. $3\frac{1}{4}$ 9. $4\frac{1}{4}$ 10. $5\frac{1}{4}$ 11. $1\frac{1}{2}$ 12. $2\frac{1}{2}$ 13. $3\frac{1}{2}$ 14. $4\frac{1}{2}$

15. $5\frac{1}{2}$ 16. $1\frac{3}{4}$ 17. $2\frac{3}{4}$ 18. $3\frac{3}{4}$ 19. $4\frac{3}{4}$ 20. $5\frac{3}{4}$

Exercise 18

How many **eighths** are there in:

1. 1   2. 2   3. 3   4. 4   5. 5   6. $1\frac{1}{2}$ 7. $2\frac{1}{2}$

8. $3\frac{1}{2}$ 9. $4\frac{1}{2}$ 10. $5\frac{1}{2}$ 11. $1\frac{1}{4}$ 12. $2\frac{1}{4}$ 13. $3\frac{1}{4}$ 14. $4\frac{1}{4}$

15. $5\frac{1}{4}$ 16. $1\frac{3}{4}$ 17. $2\frac{3}{4}$ 18. $3\frac{3}{4}$ 19. $4\frac{3}{4}$ 20. $5\frac{3}{4}$

Exercise 19

$$\frac{1}{2} = \frac{2}{4} = \frac{4}{8} = \frac{8}{16}$$

When we multiply both the top figure and the bottom figure of a fraction by **the same number** the value of the fraction **does not change.**

Copy and complete the following:

1. $\frac{2}{8} = \frac{?}{4}$ 2. $\frac{2}{5} = \frac{?}{10}$ 3. $\frac{5}{8} = \frac{10}{?}$ 4. $\frac{1}{3} = \frac{?}{9}$ 5. $\frac{6}{8} = \frac{?}{16}$

6. $\frac{1}{5} = \frac{20}{?}$ 7. $\frac{1}{4} = \frac{?}{16}$ 8. $\frac{5}{1} = \frac{?}{2}$ 9. $\frac{?}{5} = \frac{6}{15}$ 10. $\frac{3}{4} = \frac{27}{?}$

11. $\dfrac{7}{8}=\dfrac{?}{24}$ **12.** $\dfrac{5}{16}=\dfrac{10}{?}$ **13.** $\dfrac{2}{3}=\dfrac{?}{18}$ **14.** $\dfrac{3}{4}=\dfrac{?}{20}$ **15.** $\dfrac{1}{2}=\dfrac{?}{20}$

16. $\dfrac{3}{5}=\dfrac{9}{?}$ **17.** $\dfrac{?}{2}=\dfrac{10}{20}$ **18.** $\dfrac{5}{?}=\dfrac{10}{30}$ **19.** $\dfrac{3}{8}=\dfrac{27}{?}$ **20.** $\dfrac{1}{2}=\dfrac{7}{?}$

Exercise 20

To express a fraction **in its lowest terms** we divide top and bottom figures by suitable numbers until we cannot continue doing so, for example:

(a) $$\dfrac{\overset{6}{\cancel{12}}}{\underset{12}{\cancel{24}}}=\dfrac{\overset{3}{\cancel{6}}}{\underset{6}{\cancel{12}}}=\dfrac{\overset{1}{\cancel{3}}}{\underset{2}{\cancel{6}}}=\dfrac{1}{2}$$

(b) $$\dfrac{\overset{8}{\cancel{16}}}{\underset{24}{\cancel{48}}}=\dfrac{\overset{4}{\cancel{8}}}{\underset{12}{\cancel{24}}}=\dfrac{\overset{2}{\cancel{4}}}{\underset{6}{\cancel{12}}}=\dfrac{\overset{1}{\cancel{2}}}{\underset{3}{\cancel{6}}}=\dfrac{1}{3}$$

Look for **large** numbers that will divide into top and bottom figures, for example:

$$\dfrac{\overset{18}{\cancel{180}}}{\underset{24}{\cancel{240}}}=\dfrac{\overset{3}{\cancel{18}}}{\underset{4}{\cancel{24}}}=\dfrac{3}{4}$$

We first of all divide by 10 and then by 6.

Reduce the following fractions to lowest terms:

1. $\dfrac{6}{8}$ **2.** $\dfrac{24}{30}$ **3.** $\dfrac{16}{48}$ **4.** $\dfrac{3}{24}$ **5.** $\dfrac{8}{24}$ **6.** $\dfrac{3}{12}$ **7.** $\dfrac{6}{12}$

8. $\dfrac{9}{12}$ **9.** $\dfrac{2}{4}$ **10.** $\dfrac{2}{6}$ **11.** $\dfrac{3}{6}$ **12.** $\dfrac{4}{6}$ **13.** $\dfrac{2}{8}$ **14.** $\dfrac{4}{8}$

15. $\dfrac{5}{10}$ **16.** $\dfrac{6}{36}$ **17.** $\dfrac{9}{36}$ **18.** $\dfrac{12}{36}$ **19.** $\dfrac{18}{36}$ **20.** $\dfrac{24}{36}$ **21.** $\dfrac{8}{16}$

22. $\dfrac{12}{16}$ **23.** $\dfrac{14}{16}$ **24.** $\dfrac{4}{16}$ **25.** $\dfrac{6}{16}$ **26.** $\dfrac{5}{20}$ **27.** $\dfrac{10}{20}$ **28.** $\dfrac{40}{100}$

29. $\dfrac{60}{100}$ **30.** $\dfrac{2}{100}$ **31.** $\dfrac{3}{48}$ **32.** $\dfrac{4}{48}$ **33.** $\dfrac{6}{48}$ **34.** $\dfrac{8}{48}$ **35.** $\dfrac{10}{48}$

36. $\dfrac{12}{48}$ **37.** $\dfrac{16}{48}$ **38.** $\dfrac{16}{80}$ **39.** $\dfrac{30}{90}$ **40.** $\dfrac{15}{60}$

Exercise 21

An **improper fraction** is one in which the top number is larger than the bottom number, for example:

$$\frac{3}{2}, \quad \frac{8}{5}, \quad \frac{7}{2}.$$

$$\frac{3}{2} = 1\frac{1}{2}; \quad \frac{8}{5} = 1\frac{3}{5}; \quad \frac{7}{2} = 3\frac{1}{2}$$

$1\frac{1}{2}$; $1\frac{3}{5}$ and $3\frac{1}{2}$ are examples of **mixed numbers**.

Express the following **improper fractions** as **mixed numbers**.

1. $\frac{9}{4}$ 2. $\frac{8}{5}$ 3. $\frac{5}{3}$ 4. $\frac{8}{3}$ 5. $\frac{7}{3}$ 6. $\frac{10}{3}$ 7. $\frac{11}{3}$

8. $\frac{14}{3}$ 9. $\frac{5}{4}$ 10. $\frac{7}{4}$ 11. $\frac{11}{4}$ 12. $\frac{13}{4}$ 13. $\frac{17}{4}$ 14. $\frac{9}{5}$

15. $\frac{11}{5}$ 16. $\frac{12}{5}$ 17. $\frac{18}{5}$ 18. $\frac{21}{5}$ 19. $\frac{7}{6}$ 20. $\frac{13}{6}$ 21. $\frac{19}{6}$

22. $\frac{17}{6}$ 23. $\frac{22}{6}$ 24. $\frac{9}{8}$ 25. $\frac{11}{8}$ 26. $\frac{13}{8}$ 27. $\frac{19}{8}$ 28. $\frac{21}{8}$

29. $\frac{21}{10}$ 30. $\frac{31}{10}$ 31. $\frac{43}{10}$ 32. $\frac{23}{8}$ 33. $\frac{31}{8}$ 34. $\frac{51}{10}$ 35. $\frac{63}{10}$

36. $\frac{73}{10}$ 37. $\frac{27}{8}$ 38. $\frac{39}{10}$ 39. $\frac{29}{5}$ 40. $\frac{23}{4}$

Exercise 22.

To change a **mixed number** to an **improper fraction**, for example:

$$3\frac{1}{4} = \frac{17}{4}; \quad 5\frac{1}{5} = \frac{26}{5}; \quad 3\frac{1}{2} = \frac{7}{2}.$$

Change the following mixed numbers to improper fractions:

1. $3\frac{3}{4}$ 2. $2\frac{1}{4}$ 3. $1\frac{3}{4}$ 4. $4\frac{3}{4}$ 5. $5\frac{1}{4}$ 6. $6\frac{3}{4}$ 7. $7\frac{1}{2}$

8. $8\frac{1}{2}$ 9. $11\frac{1}{4}$ 10. $11\frac{1}{2}$ 11. $3\frac{1}{8}$ 12. $3\frac{3}{8}$ 13. $5\frac{1}{2}$ 14. $5\frac{3}{4}$

15. $7\frac{1}{4}$ 16. $7\frac{3}{4}$ 17. $2\frac{1}{10}$ 18. $3\frac{3}{10}$ 19. $4\frac{1}{10}$ 20. $1\frac{1}{3}$ 21. $3\frac{1}{3}$

22. $5\frac{2}{3}$ 23. $8\frac{1}{3}$ 24. $10\frac{2}{3}$ 25. $5\frac{1}{5}$ 26. $9\frac{1}{3}$ 27. $8\frac{3}{5}$ 28. $3\frac{1}{2}$

29. $11\frac{1}{2}$ **30.** $12\frac{1}{4}$ **31.** $2\frac{9}{10}$ **32.** $3\frac{1}{5}$ **33.** $7\frac{1}{8}$ **34.** $6\frac{3}{8}$ **35.** $5\frac{7}{8}$

36. $2\frac{9}{10}$ **37.** $9\frac{5}{8}$ **38.** $3\frac{7}{10}$ **39.** $1\frac{7}{8}$ **40.** $10\frac{1}{4}$

Exercise 23

Adding fractions which have **the same denominator** is easy, for example:

$$\frac{2}{5}+\frac{1}{5}=\frac{3}{5}; \quad \frac{1}{6}+\frac{5}{6}=\frac{6}{6}=1; \quad \frac{1}{3}+\frac{1}{3}=\frac{2}{3}; \quad \frac{1}{5}+\frac{4}{5}=\frac{5}{5}=1.$$

Work answers to:

1. $\frac{1}{4}+\frac{1}{4}$ **2.** $\frac{2}{3}+\frac{1}{3}+\frac{1}{3}$ **3.** $\frac{1}{10}+\frac{7}{10}$

4. $\frac{1}{8}+\frac{3}{8}+\frac{1}{8}$ **5.** $\frac{1}{2}+\frac{1}{2}+\frac{1}{2}$ **6.** $\frac{1}{5}+\frac{3}{5}+\frac{1}{5}$

7. $\frac{7}{8}+\frac{1}{8}$ **8.** $\frac{1}{10}+\frac{3}{10}$ **9.** $\frac{1}{4}+\frac{1}{4}+\frac{3}{4}$

10. $\frac{3}{8}+\frac{3}{8}+\frac{3}{8}$ **11.** $\frac{1}{8}+\frac{1}{8}+\frac{7}{8}$ **12.** $\frac{1}{5}+\frac{2}{5}+\frac{3}{5}$

13. $\frac{5}{16}+\frac{1}{16}$ **14.** $\frac{3}{16}+\frac{1}{16}+\frac{1}{16}$ **15.** $\frac{5}{16}+\frac{7}{16}$

16. $\frac{9}{16}+\frac{11}{16}$ **17.** $\frac{5}{16}+\frac{3}{16}+\frac{1}{16}$ **18.** $\frac{7}{16}+\frac{7}{16}+\frac{1}{16}$

19. $\frac{1}{10}+\frac{3}{10}+\frac{7}{10}$ **20.** $\frac{2}{5}+\frac{3}{5}+\frac{1}{5}$

Exercise 24

The easiest way to add fractions which have **different denominators** is this, for example:

(a) $\frac{1}{2}+\frac{1}{4} \; =\frac{2}{4}+\frac{1}{4} \; =\frac{3}{4}$

(b) $\frac{1}{8}+\frac{3}{4} \; =\frac{1}{8}+\frac{6}{8} \; =\frac{7}{8}$

(c) $\frac{1}{5}+\frac{1}{10}=\frac{2}{10}+\frac{1}{10}=\frac{3}{10}$

Work answers to:

1. $\dfrac{1}{2}+\dfrac{1}{5}$ 2. $\dfrac{3}{5}+\dfrac{1}{10}$ 3. $\dfrac{2}{5}+\dfrac{7}{10}$ 4. $\dfrac{1}{2}+\dfrac{1}{8}$

5. $\dfrac{3}{8}+\dfrac{1}{4}$ 6. $\dfrac{1}{2}+\dfrac{7}{8}$ 7. $\dfrac{7}{10}+\dfrac{1}{5}$ 8. $\dfrac{3}{5}+\dfrac{1}{15}$

9. $\dfrac{1}{10}+\dfrac{3}{20}$ 10. $\dfrac{3}{4}+\dfrac{7}{8}$ 11. $\dfrac{1}{4}+\dfrac{1}{4}+\dfrac{1}{8}$ 12. $\dfrac{1}{2}+\dfrac{1}{4}+\dfrac{1}{8}$

13. $\dfrac{1}{10}+\dfrac{1}{40}$ 14. $\dfrac{3}{10}+\dfrac{7}{20}$ 15. $\dfrac{5}{8}+\dfrac{1}{4}+\dfrac{1}{2}$ 16. $\dfrac{1}{5}+\dfrac{3}{10}+\dfrac{2}{5}$

17. $\dfrac{3}{8}+\dfrac{3}{4}+\dfrac{1}{2}$ 18. $\dfrac{1}{15}+\dfrac{1}{30}$ 19. $\dfrac{1}{20}+\dfrac{3}{40}$ 20. $\dfrac{1}{8}+\dfrac{3}{40}$

Exercise 25

Try now with subtraction. Here are some worked examples.

(a) $\dfrac{3}{10}-\dfrac{1}{10}=\dfrac{2}{10}=\dfrac{1}{5}$

(b) $\dfrac{3}{4}-\dfrac{1}{4}=\dfrac{2}{4}=\dfrac{1}{2}$

(c) $\dfrac{1}{5}-\dfrac{1}{10}=\dfrac{2}{10}-\dfrac{1}{10}=\dfrac{1}{10}$

(d) $\dfrac{3}{4}-\dfrac{3}{8}=\dfrac{6}{8}-\dfrac{3}{8}=\dfrac{3}{8}$

Work answers for:

1. $\dfrac{3}{4}-\dfrac{1}{4}$ 2. $\dfrac{3}{8}-\dfrac{1}{8}$ 3. $\dfrac{5}{16}-\dfrac{1}{16}$ 4. $\dfrac{2}{5}-\dfrac{1}{5}$ 5. $\dfrac{3}{10}-\dfrac{1}{10}$

6. $\dfrac{5}{8}-\dfrac{3}{8}$ 7. $\dfrac{1}{2}-\dfrac{1}{4}$ 8. $\dfrac{1}{4}-\dfrac{1}{8}$ 9. $\dfrac{1}{10}-\dfrac{1}{20}$ 10. $\dfrac{1}{4}-\dfrac{1}{16}$

11. $\dfrac{1}{8}-\dfrac{1}{16}$ 12. $\dfrac{3}{8}-\dfrac{5}{16}$ 13. $\dfrac{1}{5}-\dfrac{1}{20}$ 14. $\dfrac{3}{5}-\dfrac{1}{10}$ 15. $\dfrac{3}{4}-\dfrac{1}{12}$

16. $\dfrac{1}{8}-\dfrac{1}{24}$ 17. $\dfrac{1}{2}-\dfrac{1}{16}$ 18. $\dfrac{1}{5}-\dfrac{3}{20}$ 19. $\dfrac{3}{20}-\dfrac{1}{10}$ 20. $\dfrac{1}{4}-\dfrac{1}{10}$

Exercise 26

To add or subtract **mixed numbers**, express them first as **improper fractions**, for example:

(a) $3\frac{1}{2} + 2\frac{1}{4} = \frac{7}{2} + \frac{9}{4} = \frac{14}{4} + \frac{9}{4} = \frac{23}{4} = 5\frac{3}{4}$

(b) $5\frac{1}{8} + 2\frac{1}{2} = \frac{41}{8} + \frac{5}{2} = \frac{41}{8} + \frac{20}{8} = \frac{61}{8} = 7\frac{5}{8}$

(c) $3\frac{1}{2} - 1\frac{1}{4} = \frac{7}{2} - \frac{5}{4} = \frac{14}{4} - \frac{5}{4} = \frac{9}{4} = 2\frac{1}{4}$

Work answers to:

1. $1\frac{1}{2} + 1\frac{1}{4}$	**2.** $1\frac{1}{2} - 1\frac{1}{4}$	**3.** $2\frac{1}{2} + 3\frac{1}{4}$	**4.** $3\frac{1}{4} - 2\frac{1}{2}$	**5.** $1\frac{1}{4} + 2\frac{1}{8}$
6. $2\frac{1}{8} - 1\frac{1}{4}$	**7.** $5\frac{1}{2} + 2\frac{1}{8}$	**8.** $5\frac{1}{2} - 2\frac{1}{8}$	**9.** $3\frac{1}{2} + 4\frac{3}{4}$	**10.** $4\frac{3}{4} - 3\frac{1}{2}$
11. $7\frac{1}{2} + 3\frac{1}{4}$	**12.** $7\frac{1}{2} - 3\frac{1}{4}$	**13.** $8\frac{1}{4} + 2\frac{1}{2}$	**14.** $8\frac{1}{4} - 2\frac{1}{2}$	**15.** $6\frac{1}{4} + 2\frac{1}{8}$
16. $6\frac{1}{4} - 2\frac{1}{8}$	**17.** $3\frac{1}{4} + 2\frac{1}{2}$	**18.** $3\frac{1}{4} - 2\frac{1}{2}$	**19.** $4\frac{1}{8} + 2\frac{1}{16}$	**20.** $4\frac{1}{8} - 2\frac{1}{16}$

Exercise 27

$$\frac{1}{2} \times \frac{1}{2} = \frac{1}{4}; \quad \frac{\overset{1}{\cancel{3}}}{8} \times \frac{1}{\underset{2}{\cancel{6}}} = \frac{1}{16}; \quad \frac{3}{\underset{1}{\cancel{4}}} \times \frac{\overset{1}{\cancel{4}}}{5} = \frac{3}{5}.$$

When we **multiply** two or more fractions we first of all **do any cancelling**, as above.

Work answers to:

1. $\dfrac{1}{2} \times \dfrac{1}{4}$	**2.** $\dfrac{1}{4} \times \dfrac{3}{8}$	**3.** $\dfrac{1}{5} \times \dfrac{2}{5}$
4. $\dfrac{1}{4} \times 5$	**5.** $\dfrac{1}{2} \times 8$	**6.** $\dfrac{3}{4} \times \dfrac{1}{12}$
7. $\dfrac{1}{8} \times 16$	**8.** $\dfrac{1}{3} \times 24$	**9.** $\dfrac{2}{3} \times 5$
10. $\dfrac{1}{10} \times 40$	**11.** $\dfrac{1}{4} \times \dfrac{1}{6}$	**12.** $\dfrac{1}{4} \times \dfrac{3}{8}$
13. $\dfrac{1}{3} \times \dfrac{3}{4}$	**14.** $\dfrac{2}{3} \times \dfrac{3}{4}$	**15.** $\dfrac{2}{3} \times \dfrac{3}{8}$
16. $\dfrac{4}{5} \times \dfrac{5}{6}$	**17.** $\dfrac{4}{8} \times \dfrac{8}{9}$	**18.** $\dfrac{1}{3} \times \dfrac{3}{4}$

19. $\dfrac{1}{10} \times \dfrac{3}{10}$ **20.** $\dfrac{1}{2} \times \dfrac{4}{5}$ **21.** $\dfrac{5}{6} \times \dfrac{6}{9} \times \dfrac{3}{4}$

22. $\dfrac{3}{4} \times \dfrac{4}{5} \times \dfrac{5}{6}$ **23.** $\dfrac{5}{8} \times \dfrac{3}{5} \times \dfrac{2}{9}$ **24.** $\dfrac{2}{3} \times \dfrac{3}{8} \times \dfrac{1}{4}$

25. $\dfrac{1}{3} \times \dfrac{3}{4} \times \dfrac{4}{5}$ **26.** $\dfrac{3}{5} \times \dfrac{5}{6} \times \dfrac{6}{7}$ **27.** $\dfrac{1}{2} \times \dfrac{1}{16}$

28. $\dfrac{1}{15} \times \dfrac{3}{4}$ **29.** $\dfrac{1}{6} \times \dfrac{9}{10}$ **30.** $\dfrac{9}{10} \times \dfrac{2}{9}$

31. $\dfrac{3}{4} \times \dfrac{4}{9}$ **32.** $\dfrac{1}{6} \times \dfrac{6}{7} \times \dfrac{7}{9}$ **33.** $\dfrac{1}{10} \times 60$

34. $\dfrac{1}{5} \times 45$ **35.** $\dfrac{1}{8} \times 80$ **36.** $\dfrac{1}{2} \times \dfrac{2}{3} \times \dfrac{3}{4} \times \dfrac{4}{5}$

37. $\dfrac{1}{8} \times 100$ **38.** $\dfrac{3}{4} \times \dfrac{4}{15}$ **39.** $2 \times \dfrac{3}{4}$

40. $3 \times \dfrac{1}{3}$

Exercise 28

To multiply **mixed numbers**, express them first as **improper fractions**, for example:

(a) $3\frac{1}{2} \times 2\frac{1}{4} = \frac{7}{2} \times \frac{9}{4} = \frac{63}{8} = 7\frac{7}{8}$

(b) $1\frac{1}{4} \times 2\frac{1}{2} = \frac{5}{4} \times \frac{5}{2} = \frac{25}{8} = 3\frac{1}{8}$

(c) $1\frac{4}{5} \times 15 = \frac{9}{\cancel{5}_1} \times \frac{\cancel{15}^3}{1} = 27$

Work answers to:

1. $2\frac{1}{4} \times 2\frac{2}{3}$ **2.** $3\frac{3}{4} \times 1\frac{3}{5}$ **3.** $1\frac{5}{6} \times 3\frac{1}{2}$ **4.** $1\frac{1}{2} \times 1\frac{1}{3}$ **5.** $\frac{3}{5} \times 3\frac{1}{9}$

6. $\frac{3}{4} \times 1\frac{1}{3}$ **7.** $\frac{3}{5} \times 2\frac{1}{2}$ **8.** $1\frac{3}{4} \times 4$ **9.** $1\frac{1}{2} \times 1\frac{3}{4}$ **10.** $3 \times 3\frac{1}{3}$

11. $1\frac{1}{4} \times 1\frac{3}{5}$ **12.** $1\frac{1}{5} \times 3\frac{3}{4}$ **13.** $\frac{5}{8} \times 2\frac{2}{3}$ **14.** $\frac{3}{4} \times 2\frac{2}{5}$ **15.** $\frac{3}{5} \times 3\frac{3}{5}$

16. $1\frac{1}{2} \times 7$ **17.** $3\frac{1}{2} \times \frac{1}{2}$ **18.** $4\frac{1}{8} \times \frac{1}{4}$ **19.** $\frac{8}{9} \times 36$ **20.** $3 \times 1\frac{1}{3}$

Exercise 29

When dividing fractions or mixed numbers, turn mixed numbers to improper fractions first of all. Then turn the fraction **following** the division sign **upside down** and **multiply**, for example:

(a) $3\frac{1}{4} \div \frac{1}{2} = \frac{13}{4} \div \frac{1}{2} = \frac{13}{\cancel{4}_2} \times \frac{\cancel{2}^1}{1} = \frac{13}{2} = 6\frac{1}{2}$

(b) $4\frac{1}{8} \div 2\frac{1}{16} = \frac{33}{8} \div \frac{33}{16} = \frac{\cancel{33}^1}{\cancel{8}_1} \times \frac{\cancel{16}^2}{\cancel{33}_1} = \frac{2}{1} = 2$

(c) $5\frac{1}{2} \div \frac{3}{10} = \frac{11}{2} \div \frac{3}{10} = \frac{11}{\cancel{2}_1} \times \frac{\cancel{10}^5}{3} = \frac{55}{3} = 18\frac{1}{3}$

Work answers to:

1. $8 \div \frac{3}{4}$
2. $8 \div \frac{1}{16}$
3. $5 \div \frac{5}{6}$
4. $2 \div 1\frac{1}{2}$
5. $1\frac{1}{2} \div 2$

6. $3\frac{1}{2} \div 1\frac{3}{4}$
7. $8\frac{1}{2} \div 4\frac{1}{4}$
8. $3\frac{1}{2} \div 7$
9. $1\frac{1}{8} \div \frac{1}{4}$
10. $2 \div \frac{1}{4}$

11. $4 \div \frac{1}{3}$
12. $6 \div \frac{4}{5}$
13. $\frac{3}{8} \div 3$
14. $2\frac{4}{5} \div 1\frac{5}{6}$
15. $1\frac{1}{2} \div 2\frac{1}{4}$

16. $4\frac{2}{5} \div 1\frac{1}{10}$
17. $1\frac{2}{3} \div 2\frac{2}{9}$
18. $4\frac{1}{6} \div 3\frac{3}{4}$
19. $3\frac{1}{5} \div 1\frac{7}{25}$
20. $8\frac{1}{2} \div 4$

Problems on fractions

Example 1 Find $\frac{2}{3}$ of 18:

$$\frac{2}{\cancel{3}_1} \times \frac{\cancel{18}^6}{1} = 12$$

Example 2 Find $\frac{1}{4}$ of 80:

$$\frac{1}{\cancel{4}_1} \times \frac{\cancel{80}^{20}}{1} = 20$$

Example 3 Find $\frac{3}{8}$ of £16:

$$\frac{3}{\cancel{8}_1} \times \frac{\cancel{£16}^2}{1} = £6$$

Exercise 30

Work answers to the following:

1. Find $\frac{1}{2}$ of £4.

2. Find $\frac{3}{4}$ of £12.

3. Find $\frac{3}{4}$ of 100.

4. Find $\frac{1}{8}$ of £16.

5. Find $\frac{1}{4}$ of 20.

6. Find $\frac{3}{8}$ of 24.

7. Find $\frac{7}{8}$ of £40.

8. Find $\frac{1}{16}$ of £48.

9. Find $\frac{3}{16}$ of 32.

10. Find $\frac{1}{5}$ of £100.

11. Find $\frac{3}{5}$ of 10 kg.

12. Find $\frac{4}{5}$ of 15 kg.

13. Find $\frac{1}{10}$ of 1000.

14. Find $\frac{7}{10}$ of £500.

15. Find $\frac{3}{10}$ of 200 kg

16. Find $\frac{9}{10}$ of 1000 litres.

17. Find $\frac{1}{8}$ of £72.

18. Find $\frac{4}{5}$ of £25.

19. Find $\frac{3}{8}$ of £96.

20. Find $\frac{1}{4}$ of 1 tonne in kg.

Decimals

The number 222 = 2 hundreds + 2 tens + 2 units.

The number 222·222 = 2 hundreds + 2 tens + 2 units + 2 **tenths** + 2 **hundredths** + 2 **thousandths**.

When we 'place a point', we mark the **units column**, which lies immediately to the left of the **decimal point**.

So 42·6 means 4 tens plus 2 units plus 6 **tenths** of a unit.

Exercise 31

Write **in words** what is meant by:

1. 2·4	**2.** 3·82	**3.** 14·6	**4.** 824·1	**5.** 3·9	**6.** 28·92
7. 3·841	**8.** 6·8	**9.** 100	**10.** 100·1	**11.** 3·6	**12.** 3·72
13. 5·6	**14.** 15·82	**15.** 11·3	**16.** 2·91	**17.** 3·824	**18.** 10·6
19. 9·42	**20.** 8·123	**21.** 6·5	**22.** 3·49	**23.** 13·7	**24.** 0·642
25. 0·02	**26.** 0·008	**27.** 0·018	**28.** 3·02	**29.** 13·49	**30.** 246

In the number 32·4, ·4 is **the decimal fraction**.

To write ·4 as a vulgar fraction we write $·4 = \dfrac{4}{10}$

To write ·24 as a vulgar fraction we write $\dfrac{24}{100}$

To write ·246 as a vulgar fraction we write $\dfrac{246}{1000}$

Exercise 32

Write the following decimal fractions as vulgar fractions.

1. ·2	**2.** ·31	**3.** ·202	**4.** ·218	**5.** ·1	**6.** ·01
7. ·002	**8.** ·35	**9.** ·035	**10.** ·41	**11.** ·402	**12.** ·46
13. ·422	**14.** ·5	**15.** ·52	**16.** ·508	**17.** ·62	**18.** ·613
19. ·71	**20.** ·792				

Suppose now we wish to change a vulgar fraction, with a denominator of 10, 100, 1000 etc to a decimal fraction, for example:

Write $\dfrac{23}{1000}$ as a decimal fraction. $\dfrac{23}{1000}$ means $23 \div 1000$

Move the point **3** places to the left to divide by $1\overset{3}{\overset{\frown}{000}}$ to get ·023 ($\overset{\frown}{·\ 23·}$)

Write $\dfrac{2}{1\underset{2}{\underset{\smile}{00}}}$ as a decimal fraction. Move the point **two** places to the left to get $\dfrac{2}{100} = ·02$ ($\overset{\frown}{·02·}$)

Write $\dfrac{564}{10\,000}$ as a decimal fraction.

Move the point 4 places to the left to get $\dfrac{564}{10\,\underset{4}{\underset{\smile}{000}}} = ·0564$ ($\overset{\frown}{·\ 564·}$).

Exercise 33

Express the following vulgar fractions as decimal fractions.

1. $\dfrac{3}{10}$ 2. $\dfrac{47}{100}$ 3. $\dfrac{23}{1000}$ 4. $\dfrac{1}{10\,000}$ 5. $\dfrac{22}{100}$

6. $\dfrac{31}{1000}$ 7. $\dfrac{5}{10}$ 8. $\dfrac{29}{100}$ 9. $\dfrac{6}{10}$ 10. $\dfrac{3}{100}$

11. $\dfrac{3}{1000}$ 12. $\dfrac{5}{100}$ 13. $\dfrac{7}{1000}$ 14. $\dfrac{75}{1000}$ 15. $\dfrac{753}{10\,000}$

16. $\dfrac{35}{100}$ 17. $\dfrac{302}{10\,000}$ 18. $\dfrac{5}{1000}$ 19. $\dfrac{29}{1000}$ 20. $\dfrac{321}{10\,000}$

We are now going to add and subtract decimal numbers.

Remember to keep the decimal points **below one another**.

For example:

(a) $32·4 + 1·256 = $
$$\begin{array}{r} 32·4 \\ 1·256 \\ \hline 33·656 \end{array}$$

(b) $49·32 - 7·4 \ = $
$$\begin{array}{r} 49·32 \\ 7·4 \\ \hline 31·92 \end{array}$$

Always remember to 'pay back' after you have 'borrowed'.

Exercise 34

Work answers to:

1. $32 \cdot 9 + 1 \cdot 62$
2. $4 \cdot 84 + 29 \cdot 6$
3. $7 \cdot 248 - 1 \cdot 2$
4. $5 \cdot 6 - 2 \cdot 324$
5. $5 \cdot 81 + 2 \cdot 12 + 72 \cdot 4$
6. $18 \cdot 9 - 11 \cdot 824$
7. $56 \cdot 2 + 1 \cdot 248$
8. $0 \cdot 04 + 1 \cdot 26$
9. $1 \cdot 8 + 1 \cdot 08 + 1 \cdot 008$
10. $3 \cdot 42 - 2 \cdot 846$
11. $17 \cdot 4 - 16 \cdot 82$
12. $5 \cdot 8 + 2 \cdot 41 + 6 \cdot 328$
13. $18 \cdot 24 - 16 \cdot 99$
14. $231 \cdot 4 + 28 \cdot 91$
15. $92 \cdot 9 + 99 \cdot 98$
16. $5 \cdot 98 - 4 \cdot 99$
17. $0 + 0 \cdot 214$
18. $6 \cdot 2 + 4 \cdot 81$
19. $17 \cdot 38 - 14 \cdot 962$
20. $18 \cdot 2 - 18 \cdot 2$
21. $15 \cdot 1 + 16 \cdot 01 + 17$
22. $5 \cdot 8 + 3 \cdot 829$
23. $0 \cdot 64 + 0 \cdot 28$
24. $7 \cdot 48 - 6 \cdot 999$
25. $3 \cdot 8 + 3 \cdot 8 + 3 \cdot 8$
26. $19 \cdot 1 - 18 \cdot 1$
27. $4 \cdot 6 + 2 \cdot 8 + 1 \cdot 1$
28. $3 \cdot 12 - 2 \cdot 9$
29. $27 \cdot 9 - 26 \cdot 99$
30. $5 \cdot 6 + 5 \cdot 06 + 6 \cdot 1$
31. $100 - 98 \cdot 4$
32. $66 - 59 \cdot 83$
33. $1 \cdot 21 + 3 \cdot 6 + 4 \cdot 82$
34. $4 - 3 \cdot 999$
35. $84 + 8 \cdot 4 + 0 \cdot 84$
36. $18 \cdot 2 + 21 \cdot 4$
37. $99 \cdot 9 - 98 \cdot 89$
38. $7 \cdot 02 + 14 \cdot 84$
39. $6 \cdot 62 - 6 \cdot 619$
40. $24 \cdot 2 - 16 \cdot 1$

Multiplication and division of decimal numbers by 10, 100, 1000 etc.

When we **multiply**, the decimal point moves to the **right**.

When we **divide**, the decimal point moves to the **left**, for example:

(**a**) $32 \cdot 42 \times 10 = 32 \overset{\Large\frown}{4} \cdot 2$

            1 zero: 1 place to right

(**b**) $3 \cdot 846 \times 100 = 3 \cdot 84 \overset{\Large\frown}{6}$

            2 zeros: 2 places to right.

(**c**) $2 \cdot 5 \div 10 = 0 \cdot \overset{\Large\frown}{2} 5$

            1 zero: 1 place to left.

(**d**) $368 \div 100 = 3 \cdot \overset{\Large\frown}{68}$

            2 zeros: 2 places to left.

Remember the **whole** number 23, say, is written as 23· if we place the decimal point.

Exercise 35

Multiply:

1. 2·48 by 10
2. ·032 by 10
3. 2·481 by 100
4. ·0036 by 100
5. 48·6 by 10
6. ·0002 by 1000
7. ·372 by 100
8. ·7 by 10
9. 5·28 by 1000
10. ·00391 by 10 000
11. ·0021 by 10 000
12. 3·6 by 100
13. 8·24 by 10
14. 7·921 by 1000
15. ·0024 by 100
16. ·0001 by 1000
17. 94·6 by 100
18. ·946 by 10
19. 2·8 by 10 000
20. ·026 by 10

Exercise 36

Divide:

1. 2·78 by 10
2. 24·06 by 100
3. ·489 by 10
4. 6·92 by 100
5. 496 by 10
6. 24·1 by 1000
7. ·334 by 10
8. 2·92 by 1000
9. 4·36 by 10 000
10. 407 by 1000
11. 72 by 10 000
12. 26·42 by 100
13. 8·4 by 10
14. 9·62 by 100
15. 8·34 by 1000
16. 2·79 by 10 000
17. 4·81 by 10
18. 2·96 by 1000
19. 5·8 by 100 000
20. 1 by 1000

To multiply by a number which is **not** 10, 100, 1000 etc, for example:

(a) $3·4 \times 2 = $

$$\begin{array}{r} 3·4 \\ \times 2 \\ \hline 6·8 \end{array}$$

Multiply, as in ordinary multiplication.
We now have to place our **point**.
Count the **total** number of figures after the decimal points, in the two numbers we have just multiplied. The result is 1. Move in 1 place from the right hand side of your answer to get a result of 6·8.

(b) $4·8 \times 2·1 = $

$$\begin{array}{r} 4·8 \\ \times 2·1 \\ \hline 48 \\ 96 \\ \hline 10.08 \end{array}$$

This time move in **2** places.

(c) $\cdot48 \times 2\cdot1 =$

$$
\begin{array}{r}
\cdot48 \\
\times 2\cdot1 \\
\hline
48 \\
96 \\
\hline
1\cdot008
\end{array}
$$

This time move in **3** places.

Exercise 37

Work answers to:

1. $2\cdot4 \times 1\cdot8$	**2.** $1\cdot6 \times 2\cdot4$	**3.** $\cdot08 \times \cdot02$	**4.** $1\cdot2 \times \cdot05$
5. $\cdot46 \times \cdot1$	**6.** $\cdot36 \times \cdot01$	**7.** $\cdot28 \times 1\cdot1$	**8.** $2\cdot6 \times 1\cdot2$
9. $1\cdot5 \times 1\cdot5$	**10.** $1\cdot2 \times 1\cdot2$	**11.** $1\cdot1 \times 1\cdot2$	**12.** $5\cdot8 \times 2\cdot3$
13. $4\cdot6 \times 1\cdot1$	**14.** $9\cdot8 \times 1\cdot2$	**15.** $5\cdot5 \times 1\cdot1$	**16.** $2\cdot1 \times \cdot001$
17. $3\cdot2 \times \cdot006$	**18.** $5\cdot1 \times \cdot0002$	**19.** $3\cdot8 \times 1\cdot2$	**20.** $\cdot46 \times 1\cdot5$
21. $10\cdot1 \times 1\cdot3$	**22.** $\cdot46 \times 2$	**23.** $5\cdot9 \times 3$	**24.** $2\cdot48 \times 5$
25. $1\cdot6 \times 1\cdot1$	**26.** $3\cdot4 \times 1\cdot8$	**27.** $5\cdot1 \times 1\cdot8$	**28.** $1\cdot4 \times 1\cdot2$
29. $5\cdot2 \times 1\cdot7$	**30.** $\cdot3 \times \cdot3$	**31.** $\cdot2 \times \cdot5$	**32.** $\cdot8 \times \cdot6$
33. $\cdot8 \times \cdot06$	**34.** $\cdot08 \times \cdot6$	**35.** $\cdot04 \times 2$	**36.** $\cdot08 \times 4$
37. $\cdot008 \times 3$	**38.** $\cdot001 \times \cdot01$	**39.** $\cdot6 \times \cdot06$	**40.** $\cdot4 \times 1\cdot4$

To **divide** by a number which is not 10, 100, 1000 etc.
When we divide by a **whole** number, we need to be careful only with the position of the decimal point, for example:

(a) $3\cdot4 \div 2 \quad = 2\overline{)3\cdot4}$
$$\quad\quad\quad\quad\quad\quad 1\cdot7$$

(b) $5\cdot82 \div 3 = 3\overline{)5\cdot82}$
$$\quad\quad\quad\quad\quad\quad 1\cdot94$$

(c) $\cdot006 \div 2 = 2\overline{)\cdot006}$
$$\quad\quad\quad\quad\quad\quad \cdot003$$

We will practise by doing examples where we have to divide a decimal number by a **whole** number.

Exercise 38

Work answers to:

1. 3·62 ÷ 2	**2.** 15·8 ÷ 2	**3.** ·0056 ÷ 2	**4.** ·014 ÷ 2
5. 1·6 ÷ 4	**6.** 1·65 ÷ 5	**7.** 1·8 ÷ 2	**8.** 1·92 ÷ 3
9. ·0006 ÷ 3	**10.** 1·032 ÷ 3	**11.** 1·486 ÷ 2	**12.** 58·4 ÷ 2
13. 5·55 ÷ 5	**14.** 6·85 ÷ 5	**15.** ·005 ÷ 5	**16.** 1·00332 ÷ 3
17. 5·25 ÷ 3	**18.** 5·25 ÷ 5	**19.** 864·3 ÷ 3	**20.** 5·22 ÷ 3
21. 6·6 ÷ 6	**22.** 8·8 ÷ 4	**23.** 10·2 ÷ 2	**24.** 4·9 ÷ 7
25. 5·6 ÷ 7	**26.** 12·6 ÷ 7	**27.** 8·48 ÷ 8	**28.** 4·44 ÷ 2
29. 5·1 ÷ 3	**30.** ·0082 ÷ 2		

Percentages

Per cent means **per hundred**.

So 5 per cent (written 5%) means 5 hundredths i.e. $\frac{5}{100}$.

Giving this in its lowest terms we get 5% $= \frac{1}{20}$.

Exercise 39

Give each of the following percentages as fractions in their lowest terms.

1. 1%	2. 2%	3. 4%	4. 6%	5. 8%	6. 10%
7. 12%	8. 14%	9. 15%	10. 16%	11. 18%	12. 20%
13. 22%	14. 24%	15. 25%	16. 26%	17. 28%	18. 30%
19. 32%	20. 34%	21. 35%	22. 36%	23. 38%	24. 40%
25. 42%	26. 44%	27. 45%	28. 46%	29. 48%	30. 50%
31. 52%	32. 54%	33. 55%	34. 56%	35. 58%	36. 60%
37. 62%	38. 64%	39. 65%	40. 66%	41. 68%	42. 70%
43. 72%	44. 74%	45. 75%	46. 76%	47. 78%	48. 80%
49. 82%	50. 84%	51. 85%	52. 86%	53. 88%	54. 90%
55. 92%	56. 94%	57. 95%	58. 96%	59. 98%	60. 100%
61. 150%	62. 200%	63. 250%	64. 300%	65. 350%	66. 400%
67. 450%	68. 110%	69. 120%	70. 160%	71. 210%	72. 220%
73. 280%	74. 340%	75. 310%	76. 320%	77. 500%	78. 550%
79. 290%	80. 275%				

Problems in percentages

Example 1: Find 50% of £10.

$$50\% = \frac{1}{2} \text{ and } \frac{1}{2} \text{ of } £10 = £5.$$

So 50% of £10 = £5.

Example 2:   Find 20% of £300.

$$20\% = \frac{1}{5} \text{ and} \frac{1}{5} \text{ of £300} = £60.$$

So 20% of £300 = £60.

Example 3:   Find 10% of 80 metres.

$$10\% = \frac{1}{10} \text{ and} \frac{1}{10} \text{ of 80 metres} = 8 \text{ metres}.$$

So 10% of 80 metres = 8 metres.

Exercise 40

Work answers to:

1. 10% of £400	2. 20% of 10 tonnes	3. 25% of £20
4. 250% of 12	5. 5% of £80	6. 110% of £70
7. 20% of 45 cm	8. 4% of 50 cm	9. 10% of 500 cm³
10. 75% of £4	11. 200% of £100	12. 300% of 6 metres
13. 5% of 1 million	14. 30% of £100	15. 50% of 2 km
16. 25% of 8 km	17. 75% of 8 kg	18. 70% of £1000
19. 8% of £5000	20. 130% of £5	

Profit and Loss

Write cost price as C.P.
Write selling price as S.P.
If the selling price is bigger than the cost price **a gain has been made**.
If the cost price is bigger than the selling price **a loss has been made**.

Example 1: C.P. = £4; S.P. = £5; Gain = £1.

Example 2: C.P. = £4; S.P. = £3: Loss = £1.

Exercise 41

Copy and complete the following table.
In each case **1 article** is being bought and sold.

Cost Price	Selling Price	Gain	Loss
£3	£2		
£4	£5		
£1	£1·50		
£2	£3		
£8	£8·50		
£10	£12		
£7	£7·25		
£6	£4·80		
£8	£8·10		
£6	£3·50		
£18	£23·50		
£9	£10·20		
£11	£11·20		
£8·50	£7		
£6·25	£3·20		

Can you say why only the gain **or** loss column will have an entry for each question?

We must always compare the cost price and selling price of **the same number of articles** before we can decide on a gain or a loss.

Example 1: 3 articles are bought at a cost of £2 each. 2 are sold for £3 each and the third is sold for £1. Work out the **total gain** or the **total loss**.

C.P. of 3 articles = 3 × £2 = £6

S.P. of 3 articles = 2 × £3 + £1
$$= £6 + £1$$
$$= £7$$

So total gain is £1.

Exercise 42

Copy and complete the table. In each case, **all** the articles were bought and sold.

No. Bought	Cost Price of 1	Selling Price of 1	Total Gain	Total Loss
7	£2	£2·30		
2	£1	£1·10		
8	£8	£7·59		
10	£3·50	£3·60		
6	£2·40	£1·80		
4	£5	£3·80		
24	£6	£6·20		
40	£4·20	£3·80		
100	£6·40	£5·80		
20	£2·30	£2·40		
200	£7·10	£7·40		
12	£2·20	£2·10		
10	£1·80	£1·50		
80	£3·80	£4·50		
6	£2	£2·01		

Simple Interest

When money is borrowed **interest** has to be paid. Usually, money is lent at a certain **rate of interest** and for **a fixed time**. We use a formula to calculate interest. We call the interest **simple interest** when the interest stays the same each year. **'Per annum'** means **'each year'**.

Example 1: What interest is due on £400 borrowed for 3 years at 10% per annum simple interest?

We use the formula $I = \dfrac{P \times R \times T}{100}$ where I is the **interest**, P is the **principal** and T is the **time**.

I is in £s; P is in £s; T is in years.

$$I = \frac{P \times R \times T}{100} = £\frac{400 \times 10 \times 3}{100} = £120$$

So, simple interest = £120.

Example 2: What interest would be due on £1000 borrowed for 5 years at 20% per annum simple interest?

$$I = \frac{P \times R \times T}{100} = £\frac{1000 \times 20 \times 5}{100} = £1000$$

So, simple interest is £1000.

Exercise 43

Copy and complete the following table.

	Sum borrowed	Rate per annum	Time	Simple Interest
1.	£200	5%	2 years	
2.	£500	6%	3 years	
3.	£1000	10%	2 years	
4.	£800	5%	3 years	
5.	£900	20%	2 years	
6.	£600	15%	1 year	
7.	£1500	12%	3 years	
8.	£2000	10%	2 years	
9.	£500	5%	5 years	
10.	£800	20%	1 year	
11.	£3000	10%	2 years	
12.	£4000	8%	3 years	
13.	£5000	10%	2 years	
14.	£200	5%	8 years	
15.	£400	10%	10 years	

Foreign Money

When we go abroad we change our British money into the **currency** of the country to which we travel.

Here are a few names of **foreign currency** and the name of the country to which each belongs.

France: Franc

Italy: Lire

Spain: Peseta

Germany: Deutschmark

We only need to be able to multiply or divide to work out foreign currency problems.

Example 1:   An Indian rupee is worth 28 pence. How many rupees do I get for £2·80?

$$\text{Number of rupees} = \frac{£2·80}{28p} = \frac{280p}{28p} = 10.$$

Example 2:   20 German marks are worth £1 sterling. How many pounds sterling do I get for 120 German marks?

$$\text{Number of pounds sterling} = \frac{120}{20} = £6.$$

Copy and complete the following table. The **exchange rates** vary from day to day.

Exercise 44

	Currency in hand	Currency Required	Exchange Rate	Amount Received
1.	£4	German Marks	21 marks to £1	
2.	94 Rupees	Pounds Sterling	1 Rupee = 18p	
3.	270 Francs	Pounds Sterling	90 Francs = £1	
4.	4800 Dollars	Pounds Sterling	2·4 Dollars = £1	
5.	£36	Rupees	1 Rupee = 18p	
6.	4800 Lira	Pounds Sterling	1200 Lira = £1	
7.	£5·50	Pesetas	140 Pesetas = £1	
8.	£40	Francs	92 Francs = £1	
9.	156 Dollars	Pounds Sterling	2·6 Dollars = £1	
10.	2400 Francs	Pounds Sterling	100 Francs = £1	
11.	£44·60	Rupees	1 Rupee = 20p	
12.	17 000 Lira	Pounds Sterling	1000 Lira = £1	
13.	26 000 Pesetas	Pounds Sterling	130 Pesetas = £1	
14.	880 Marks	Pounds Sterling	22 Marks = £1	
15.	£85	Dollars	2·4 Dollars = £1	

Hire Purchase

When we buy goods on hire purchase we **spread the payments** over a period of time. Usually a **deposit** is paid before the goods can be taken from the shop. Weekly or monthly instalments are paid and **extra** is charged for hire purchase. In each case, calculate the **extra** charge for hire purchase. Copy and complete the table.

Exercise 45

	Cash Price of Goods	Deposit	Number of Payments	Amount of 1 payment	Extra charged for H.P.
1.	£80	£15	12	£6·20	
2.	£180	£30	24	£7·30	
3.	£300	£50	36	£7·50	
4.	£1000	£200	60	£16·50	
5.	£150	£40	12	£11·20	
6.	£480	£120	36	£12·50	
7.	£320	£100	12	£24·50	
8.	£460	£140	36	£10·80	
9.	£4000	£1000	60	£62·50	
10.	£5000	£1500	60	£65·50	
11.	£90	£20	12	£6·90	
12.	£120	£30	24	£5·25	

36

How Your Wage is Calculated

Suppose you have a basic working week of 40 hours. For this you will receive a **guaranteed** weekly wage. If you work overtime you may get a higher rate of pay.

Example 1: A man works a basic 40 hour week for a guaranteed wage of £80. In a certain week he works 10 hours overtime, 6 hours at **time and half** and the rest at **double time**.
Calculate his **gross weekly wage**.
40 hours for £80 means his basic rate is £2 per hour.
Time and half earns him **£3** per hour.
Double time earns him **£4** per hour.
Gross weekly wage is £80 + 6 × £3 + 4 × £4 = £80 + £18 + £16 = £114.
Gross weekly wage is the amount earned **before** deductions for income tax, national insurance etc. What you actually receive is your **net** wage.
Copy and complete the wages table.

Exercise 46

	Basic	Actual time worked	Basic rate per hour	Hours at time and half	Hours at double time	Gross weekly wage
1.	40 hours	50 hours	£2	2		
2.	38 hours	46 hours	£3		3	
3.	36 hours	46 hours	£4	5		
4.	40 hours	42 hours	£3		1	
5.	30 hours	38 hours	£2		4	
6.	36 hours	42 hours	£4		4	
7.	42 hours	52 hours	£2		8	
8.	38 hours	49 hours	£3		7	
9.	40 hours	46 hours	£4	2		
10.	42 hours	48 hours	£6		3	
11.	38 hours	46 hours	£3		2	
12.	40 hours	52 hours	£3		4	
13.	42 hours	52 hours	£2		5	
14.	34 hours	46 hours	£3	3		
15.	28 hours	40 hours	£4	6		

Suppose you walk for 4 hours at a speed of 3 km/hour. You will have covered a distance of 12 km. We say that Distance = Speed × Time.

Example 1: How far do you travel at 40 km/h in 3 hours?
$$D = S \times T = 40 \times 3 \text{ km} = 120 \text{ km}.$$

Example 2: How far do you travel at 80 km/h in 10 hours?
$$D = S \times T = 80 \times 10 \text{ km} = 800 \text{ km}.$$

Example 3: How far do you travel at 100 km/h in ½ hour?
$$D = S \times T = 100 \times \text{½ km} = 50 \text{ km}.$$

Exercise 47

A car is travelling at an average speed of 40 km/hour. What distance will be travelled in:

1. 2 hours
2. 3 hours
3. $\frac{1}{2}$ hour
4. 15 minutes
5. 6 hours
6. 10 hours
7. $3\frac{1}{2}$ hours
8. 2 hours 20 minutes
9. $4\frac{1}{2}$ hours
10. 8 hours 45 minutes

If we know the **distance** and the **speed** we can find the **time taken**, for example:

A man walks a distance of 20 km at a speed of 4 km/hour. How long does this take?

We say that Time $= \dfrac{\text{Distance}}{\text{Speed}}$: $T = \dfrac{D}{S}$

So time taken $= \dfrac{20 \text{ km}}{4 \text{ km/hour}} = 5 \text{ hours}$

Example: How long does it take to walk 9 km at a speed of 3 km/h?
$$T = \frac{D}{S} = \frac{9}{3} = 3 \text{ hours}.$$

Exercise 48

A distance of 200 km is to be travelled. How long does this take at:

1. 40 km/hr
2. 20 km/hr
3. 50 km/hr

4.	10 km/hr	5.	100 km/hr	6.	60 km/hr
7.	80 km/hr	8.	25 km/hr	9.	75 km/hr
10.	30 km/hr				

Now think about this.

In 8 hours you travel a distance of 96 km. So in 1 hour you travel 12 km. We say that your speed is 12 km/hour.

So we have Speed $=\dfrac{\text{Distance}}{\text{Time}}$: $S = \dfrac{D}{T}$; for example

(a) What speed will you travel at to cover 100 km in 2 hours?

$S = \dfrac{D}{T} = \dfrac{100}{2}$ km/hr = 50 km/hr.

(b) What speed will you travel at to cover 200 km in 10 hours?

$S = \dfrac{D}{T} = \dfrac{200}{10} = 20$ km/hr.

Copy and complete the table.

Exercise 49

	Distance	Speed	Time
1.	80 km	20 km/hr	
2.	40 km	10 km/hr	
3.		20 km/hr	4 hours
4.	400 km		8 hours
5.	90 km	30 km/hr	
6.	84 km		7 hours
7.		50 km/hr	8 hours
8.	400 km	40 km/hr	
9.	10 000 km		10 hours
10.	50 km	5 km/hr	
11.		40 km/hr	8 hours
12.	100 km		5 hours
13.	900 km	90 km/hr	
14.	200 km	50 km/hr	
15.	300 km		30 hours
16.		18 km/hr	2 hours

The Metric System: (Length)

Units in general use are metres, kilometres and centimetres.
100 cm = 1 m   large unit → small unit; multiply.
1000 m = 1 km   small unit → large unit; divide.

Exercise 50

1. How many centimetres are there in 2 metres?

2. How many centimetres are there in 3·5 metres?

3. How many centimetres are there in 10 metres?

4. How many centimetres are there in 5 metres?

5. How many centimetres are there in 2·4 metres?

6. How many metres are there in 2 kilometres?

7. How many metres are there in 10 kilometres?

8. How many metres are there in 15 kilometres?

9. How many metres are there in 20 kilometres?

10. How many metres are there in 22 kilometres?

11. Change 200 centimetres to metres.

12. Change 400 centimetres to metres.

13. Change 4000 centimetres to metres.

14. Change 8000 centimetres to metres.

15. Change 80 000 centimetres to metres.

16. Change 900 centimetres to metres.

17. Change 2000 metres to kilometres.

18. Change 10 000 metres to kilometres.

19. Change 5000 metres to kilometres.

20. Change 3500 metres to kilometres.

The Metric System: (Area)

A rectangle has a length of 6 cm and a width of 4 cm. Calculate its area in cm^2.

Area = L × B
     = 6 cm × 4 cm
     = 24 cm^2.

To find the length if we know A and B use $L = \dfrac{A}{B}$

To find the breadth if we know A and L use $B = \dfrac{A}{L}$

$cm \times cm = cm^2$.
$metres \times metres = metres^2$.

Copy and complete the table.

Exercise 51

	Length	Breadth	Area
1.	6 cm	5 cm	
2.	4 cm	3 cm	
3.		2 cm	12 cm^2
4.	8 m	2 m	
5.		2 m	10 m^2
6.	4 m	$2\frac{1}{2}$ m	
7.		$2\frac{1}{2}$ m	20 m^2
8.	3 cm	$1\frac{1}{2}$ cm	
9.	18 cm	10 cm	
10.		5 m	40 m^2
11.	4 m	3 m	
12.		5 cm	60 cm^2
13.	16 cm		80 cm^2
14.	2 m	1 m	
15.	$3\frac{1}{2}$ m	2 m	
16.	30 m		90 m^2
17.	6 cm	$5\frac{1}{2}$ cm	
18.		$1\frac{1}{2}$ cm	$4\frac{1}{2}$ cm^2
19.	$8\frac{1}{2}$ m	4 m	
20.	6 m	$3\frac{1}{2}$ m	

For **larger** areas we use **hectares**.
1 hectare = 10 000 metres^2, for example:

A field is 200 metres long and 150 metres wide. Calculate its area in **hectares**.

Area = L × B = 200 m × 150 m
= 30 000 m^2
= $\dfrac{30\,000}{10\,000}$ hectares = 3 **hectares**.

Exercise 52

Copy and complete the table.

	Length	Breadth	Area in Hectares
1.	300 m	200 m	
2.	400 m	200 m	
3.	500 m	100 m	
4.	1500 m	1000 m	
5.	800 m	700 m	
6.	450 m	300 m	
7.	650 m	200 m	
8.	320 m	40 m	
9.	380 m	200 m	
10.	1000 m	800 m	

The Metric System: (Volume)

Volume of a cuboid = L × B × H.
A cube's **length, breadth** and **height** are equal.

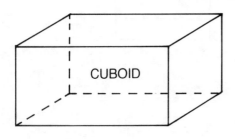

CUBOID

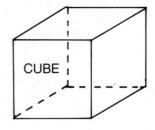

CUBE

42

Volumes are measured in **cubic units**, for example:

A cuboid is 6 cm long, 4 cm wide and 2 cm high. Find its volume.
$$V = l \times b \times h = 6\,cm \times 4\,cm \times 2\,cm$$
$$= 48\,cm^3$$

Exercise 53

Copy and complete the table.

	Length	Breadth	Height	Volume
1.	3 cm	2 cm	1 cm	cm^3
2.	8 cm	6 cm	4 cm	cm^3
3.	1 m	20 m	10 m	m^3
4.	1·5 m	1 m	1 m	m^3
5.	20 cm	15 cm	10 cm	cm^3
6.	$\frac{1}{2}$ m	20 cm	20 cm	cm^3
7.	$\frac{1}{2}$ m	$\frac{1}{4}$ m	$\frac{1}{8}$ m	m^3
8.	3 m	2 m	1 m	m^3
9.	200 cm	100 cm	50 cm	m^3
10.	300 cm	200 cm	100 cm	m^3
11.	1 m	1 m	1 m	m^3
12.	10 cm	10 cm	10 cm	cm^3
13.	20 cm	5 cm	2 cm	cm^3
14.	4 m	2 m	1 m	m^3
15.	10 m	8 m	500 cm	m^3

Capacity

1 litre $= 1000\,cm^3$.

We say that a container 'holds' a liquid. Suppose the container has **an inside volume** of 2000 cm^3. We say that the **capacity** of the container is $\dfrac{2000}{1000}$ litres i.e. 2 litres.

How many litres do the following cuboids hold if the inside measurements are as follows:

Exercise 54

The first one is worked for you.

	Length	Breadth	Height	Capacity
1.	40 cm	30 cm	20 cm	$\dfrac{40 \times 30 \times 20}{1000}$ litres = 24 litres
2.	50 cm	50 cm	40 cm	
3.	40 cm	40 cm	20 cm	
4.	100 cm	90 cm	80 cm	
5.	80 cm	80 cm	60 cm	
6.	50 cm	30 cm	30 cm	
7.	10 cm	10 cm	10 cm	
8.	100 cm	60 cm	20 cm	
9.	45 cm	20 cm	10 cm	
10.	95 cm	60 cm	20 cm	
11.	25 cm	20 cm	10 cm	
12.	12 cm	10 cm	5 cm	

We all know what a 1 kilogram bag of sugar looks like and we know how heavy it feels. There are 1000 grams in 1 kilogram. There are 1000 kilograms in **1 tonne**.

Large unit → small unit; multiply.
Small unit → large unit; divide.

Exercise 55

Work answers to the following:

1. Change 2 kg to grams.
2. Change 2 tonnes to kg.
3. Change 5940 g to kg.
4. Change 826 kg to tonnes
5. Change 426 g to kg.
6. Change 4·5 tonnes to kg.
7. Change 3·2 tonnes to kg.
8. Change 2000 g to kg.
9. Change 3500 kg to tonnes.
10. Change 8 kg to grams.
11. Chanage 2·48 kg to grams.
12. Chanage 3·8 tonnes to kg.
13. Change 1·2 kg to grams.
14. Change 10·2 tonnes to kg.
15. Change 32 469 g to kg.
16. Change 8·4 kg to grams.
17. Change 5 tonnes to kg.
18. Change 1·2 tonnes to grams.
19. Change 0·5 tonnes to grams.
20. Change 18·6 kg to grams.

Basic Algebra

$+, -, \times$ and $\div$ are used in algebra in the same way as we use them in arithmetic. In algebra we use **letters** to take the place of numbers. A letter can stand for any number.

$a + b$ means *add a and b*.

Once we know what the letter stands for we just have an **arithmetic** problem.

Example 1:   What is the value of $a + b$ if $a = 2$ and $b = 3$?
Simply 'plug in' 2 for a and 3 for b to get $a + b$.
$a + b = 2 + 3 = 5$.

Example 2:   If $a = 5$ and $b = 7$ what is the value of $a + b$?
Again we have $a + b = 5 + 7 = 12$.

Now try the following exercise:

Exercise 1

Work answers for $a + b$ when:

1. $a = 2; b = 1$	2. $a = 1; b = 4$	3. $a = 1; b = 3$
4. $a = 7; b = 4$	5. $a = 3; b = 17$	6. $a = 5; b = 13$
7. $a = 21; b = 17$	8. $a = 7; b = 11$	9. $a = 3; b = 19$
10. $a = 12; b = 14$	11. $a = 17; b = 13$	12. $a = 2; b = 21$
13. $a = 11; b = 21$	14. $a = 1; b = 18$	15. $a = 2; b = 26$
16. $a = 1; b = 27$	17. $a = 9; b = 8$	18. $a = 16; b = 13$
19. $a = 49; b = 1$	20. $a = 48; b = 6$	21. $a = 5; b = 106$
22. $a = 14; b = 32$	23. $a = 7; b = 2\frac{1}{2}$	24. $a = 100; b = 6$
25. $a = 12; b = 47$	26. $a = 17; b = 8\frac{1}{2}$	27. $a = 102; b = 14$
28. $a = 17; b = 23$	29. $a = 1; b = 60$	30. $a = 4\frac{1}{2}; b = 2\frac{1}{2}$
31. $a = 10; b = 100$	32. $a = 41; b = 61$	33. $a = \frac{1}{4}; b = \frac{3}{4}$
34. $a = \frac{1}{8}; b = \frac{7}{8}$	35. $a = 22; b = 77$	36. $a = 4; b = 600$
37. $a = 1; b = 132$	38. $a = \frac{1}{10}; b = \frac{7}{10}$	39. $a = 2; b = 2\frac{1}{4}$
40. $a = 100; b = 1000$		

Exercise 2

Work answers for $x - y$ when:

1. $x = 5; y = 1$
2. $x = 4; y = 1$
3. $x = 3; y = 2$
4. $x = 5; y = 4$
5. $x = 19; y = 1$
6. $x = 17; y = 2$
7. $x = 26; y = 2$
8. $x = 24; y = 19$
9. $x = 25; y = 5$
10. $x = 28; y = 3$
11. $x = 30; y = 20$
12. $x = 32; y = 1$
13. $x = 14; y = 12$
14. $x = 15; y = 6$
15. $x = 46; y = 23$
16. $x = 45; y = 23$
17. $x = 52; y = 26$
18. $x = 11; y = 7$
19. $x = 10; y = 6$
20. $x = 9; y = 7$
21. $x = 4; y = 1\frac{1}{2}$
22. $x = 2; y = \frac{1}{4}$
23. $x = 1; y = \frac{1}{2}$
24. $x = 100; y = 24$
25. $x = 1000; y = 100$
26. $x = 23; y = 17$
27. $x = 10; y = 6\frac{1}{2}$
28. $x = 12; y = 9\frac{1}{2}$
29. $x = 3; y = \frac{1}{2}$
30. $x = 10; y = 1\frac{1}{4}$
31. $x = 100; y = 27$
32. $x = 25; y = 24\frac{1}{2}$
33. $x = 1; y = \frac{1}{4}$
34. $y = 40; y = 39$
35. $x = 37\frac{1}{2}; y = \frac{1}{2}$
36. $x = 10\frac{1}{2}; y = 1\frac{1}{2}$
37. $x = 14; y = 7$
38. $x = 12; y = 6$
39. $x = 4; y = 0$
40. $x = 6; y = 0$

When we write $3y$ we mean 3 **times** y.
We could write $3y = 3 \times y$.
When $y = 2; 3y = 3 \times 2 = 6$.
When $y = 5; 3y = 3 \times 5 = 15$.
When $y = 10; 3y = 3 \times 10 = 30$.

Exercise 3

Work answers to the following when $a = 4$.

1. $2a$
2. $3a$
3. $6a$
4. $5a$
5. $7a$
6. $\frac{1}{2}a$
7. $\frac{1}{4}a$
8. $8a$
9. $\frac{3}{4}a$
10. $10a$
11. $9a$
12. $13a$
13. $11a$
14. $12a$
15. $\frac{1}{8}a$
16. $20a$
17. $\frac{1}{5}a$
18. $\frac{1}{10}a$
19. $30a$
20. $100a$
21. $a + 6$
22. $2a - 1$
23. $a + 8$
24. $3a - 4$
25. $2a + 7$
26. $\frac{1}{2}a + 1$
27. $\frac{1}{4}a + 3$
28. $13a - 50$

29. $12a - 47$ **30.** $5a + 2$ **31.** $3a + 19$ **32.** $4\frac{1}{2}a + 1$

33. $7\frac{1}{2}a - 10$ **34.** $8\frac{1}{2}a - 4$ **35.** $15a - 58$ **36.** $13a + 1$

37. $50a - 200$ **38.** $3a + 8$ **39.** $2a - 4$ **40.** $\frac{1}{2}a - 2$

Example 1: If $a = 2$ and $c = 3$ work an answer for $3a + 2c$.
$$3a + 2c = 3 \times 2 + 2 \times 3 = 6 + 6 = 12$$

Example 2: If $x = 1$ and $y = 3$ work an answer for $2x + 3y$.
$$2x + 3y = 2 \times 1 + 3 \times 3 = 2 + 9 = 11$$

Remember to **multiply** before you **add**.

Exercise 4

If $t = 2$ and $p = 3$ work answers to:

1. $2t + p$ **2.** $p + 3t$ **3.** $2t - p$ **4.** $5t + p$

5. $6t + 3p$ **6.** $4t - p$ **7.** $t + 2p$ **8.** $t + 3p$

9. $4t + p$ **10.** $3t + p$ **11.** $3t + 5p$ **12.** $t + 7p$

13. $7t + p$ **14.** $8t - p$ **15.** $6t - p$ **16.** $8t + 2p$

17. $8t - 2p$ **18.** $3t + 7p$ **19.** $15t + p$ **20.** $15t - 4p$

21. $3p - t$ **22.** $p + 8t$ **23.** $\frac{1}{2}t + p$ **24.** $8p + 10t$

25. $7p - 20$ **26.** $p - 3$ **27.** $8p - 24$ **28.** $\frac{1}{2}t - 1$

29. $\frac{1}{2}t + 1$ **30.** $5t + 15p$ **31.** $10p$ **32.** $6p - 9t$

33. $5p - 14$ **34.** $10p + 1$ **35.** $\dfrac{t + p}{5}$ **36.** $\dfrac{p - t}{2}$

37. $4p$ **38.** $16p + 1$ **39.** $8t - 15$ **40.** $3t - 5$

When we write ab we mean $a \times b$. $a \times b$ has the same meaning as $b \times a$.

Example 1: If $a = 4$ and $b = 2$ what is the value of $3ab$?
$$3ab = 3 \times a \times b = 3 \times 4 \times 2 = 24.$$

Example 2: If $a = 1$ and $b = 6$ what is the value of $4ab$?
$$4ab = 4 \times 1 \times 6 = 24.$$

Exercise 5

If $c = 2$ and $d = 3$ work answers to:

1. $2cd$ **2.** $2dc$ **3.** $3cd$ **4.** $3dc$ **5.** $4cd$

6. $\frac{1}{2}cd$ 7. $\frac{1}{3}cd$ 8. $8cd$ 9. $9cd$ 10. $1\frac{1}{2}cd$

11. $4\frac{1}{2}cd$ 12. $5\frac{1}{2}cd$ 13. $10cd$ 14. $20cd$ 15. $\frac{1}{6}cd$

16. $5cd$ 17. $2\frac{1}{2}cd$ 18. $40cd$ 19. $100cd$ 20. $50cd$

Exercise 6

If $a = 2$; $b = 3$; $c = 4$ and $d = 6$ work answers to:

1. $2a + b$ 2. $c + 3b$ 3. $d + 2c$ 4. $a + 2b$ 5. $c + 3d$

6. $b - a$ 7. $5d - c$ 8. $3c + a$ 9. $5c + d$ 10. $5a - b$

11. $4c + 3d$ 12. $7b + c$ 13. $21a + b$ 14. $14a - d$ 15. $3b - a$

16. $10a + b$ 17. $11c - d$ 18. $12d - a$ 19. $6b + 5c$ 20. $6d + 5a$

Exercise 7

If $a = 5$ and $b = 3$ work answers to:

1. $\dfrac{3a}{b}$ 2. $\dfrac{5b}{a}$ 3. $\dfrac{a + b}{8}$ 4. $\dfrac{a - b}{2}$ 5. $\dfrac{6a}{b}$

6. $\dfrac{2b}{6}$ 7. $\dfrac{8a}{20}$ 8. $\dfrac{4b}{6}$ 9. $\dfrac{20a}{50}$ 10. $\dfrac{4a}{5}$

11. $\dfrac{7b}{3}$ 12. $\dfrac{10b}{6}$ 13. $\dfrac{2ab}{15}$ 14. $\dfrac{5ab}{25}$ 15. $\dfrac{6ab}{45}$

16. $\dfrac{12a}{5}$ 17. $\dfrac{12b}{9}$ 18. $\dfrac{12a}{15}$ 19. $\dfrac{2a + b}{13}$ 20. $\dfrac{3b + a}{7}$

Exercise 8

If $a = 1$; $b = 2$ and $c = 3$ find answers to:

1. $a + b + c$ 2. $a + b - c$ 3. $2a + b + c$

4. $2a + b - c$ 5. $3a + b + c$ 6. $3a + b - c$

7. $3a + 2b + c$ 8. $3a + 3b - c$ 9. $3a + 2b + 3c$

10. $2a + 3b + 3c$ 11. $a + 5b - c$ 12. $a + 6b + 2c$

13. $5a + b + 8c$ 14. $6a + b + c$ 15. $6a + b - c$

16. $10a + 2b + 5c$ 17. $10a - 2b + c$ 18. $5a + 5b + c$

19. $5a + 3b - c$ 20. $3(a + b + c)$

Exercise 9

If $x = 2$; $y = 3$ and $t = 8$ find answers for:

1. $x \times t$	**2.** $y + t$	**3.** $t \div x$	**4.** $2yt$
5. ytx	**6.** $2x + t$	**7.** $x + yt$	**8.** $t + 3y$
9. $x + 4y$	**10.** $y - x$	**11.** $t - x$	**12.** $t - y$
13. $5t$	**14.** $8y$	**15.** $7x + y$	**16.** $10x$
17. $t + 8x$	**18.** $2(x + y)$	**19.** $3(y + t)$	**20.** $y + t + 3x$

Exercise 10

If $t = 2$; $x = 4$ and $m = 6$ work answers for:

1. $\dfrac{t}{2}$	**2.** $\dfrac{x}{2}$	**3.** $\dfrac{m}{2}$	**4.** $\dfrac{2t}{4}$
5. $3m$	**6.** $m \div x$	**7.** $2x \div m$	**8.** $\dfrac{t + m}{4}$
9. $\dfrac{m + x}{5}$	**10.** $\dfrac{m + 2t}{5}$	**11.** mxt	**12.** $mx \div t$
13. $\dfrac{t}{10}$	**14.** $mt + 3x$	**15.** $xt + \dfrac{m}{2}$	**16.** $\dfrac{x}{2} + m$
17. $2x + 3m$	**18.** $x + 4m + t$	**19.** $3t + 4m$	**20.** $m + 4x - t$

Exercise 11

If $x = 5$ and $y = 6$ work answers to:

1. $2(x + y)$	**2.** $\dfrac{x +}{11}$	**3.** $\dfrac{2x}{5}$	**4.** $\dfrac{y}{2}$
5. $\dfrac{3y + x}{23}$	**6.** $\dfrac{1}{2}(2x + y)$	**7.** $x + 2y$	**8.** $x + 3y$
9. $y - x$	**10.** $3(y - x)$	**11.** $4(2x - y)$	**12.** $3(3x - y)$
13. $\dfrac{2x - y}{8}$	**14.** $\dfrac{5y}{2}$	**15.** $\dfrac{8y}{3}$	**16.** $5x$
17. $7y$	**18.** $\dfrac{5x - y}{19}$	**19.** $11x$	**20.** $\dfrac{y}{12}$

50

Exercise 12

Write down, using letters and signs, the following instructions.

1. add p and t. Divide your answer by x.

2. Take p from q.

3. Take twice m from c.

4. Take a from b and divide the result by p.

5. Add a, b and t.

6. Add twice t to four times q.

7. Add t and w and multiply the result by v.

8. Subtract m from twice t.

9. Multiply c by d and subtract 5 from your result.

10. Divide p by q.

11. Divide the sum of p and q by t.

12. Divide the sum of x and y by their difference, if x is greater than y.

13. Take twice q from four times d.

14. Add five times t to three times m.

15. Divide twice r by three times p.

16. Multiply the sum of x and r by t.

17. Multiply t by q. Add 3 to your answer.

18. Divide p by q. Subtract 5 from your answer.

19. Add a, b and c. Multiply your answer by m.

20. Add a and b and subtract x from their sum.

Like and Unlike Terms

$2x$ and $5x$ are **like** terms.
$3a$ and $8a$ are **like** terms.
$2m$ and $11m$ are **like** terms.
$5p$ and $2q$ are **unlike** terms.
$2a$ and $3c$ are **unlike** terms.
$5m$ and $3x$ are **unlike** terms.

Like terms can be added or subtracted to give **simpler results**, for example:

(a) $5x + 7x = 12x$

(b) $13x - 4x = 9x$

(c) $2x + 11x - 3x = 10x$

(d) $5y + 7y - 3y = 9y$

Unlike terms cannot be added or subtracted to give simpler results, for example:

(a) $5p + 3q$ must **remain** as $5p + 3q$.

(b) $2m + 7x$ must **remain** as $2m + 7x$.

Example 1:   Find a simpler answer for $3a + c + 4a + 7c$.
Simpler answer is $7a + 8c$.

Example 2:   Find a simpler answer for $2x + y + 11x + 4y + x + 3y$.
Simpler answer is $14x + 8y$.

Exercise 13

Give simpler answers for:

1. $x + 3x$

2. $3y + 11y$

3. $5t + 3t$

4. $x + 7x + 2x$

5. $3a + 11a + 2a$

6. $5x + 8x + 3x$

7. $4c + 8c + c$

8. $x + 3x + 5x + 2x$

9. $y + 4y + 7y + 2y$

10. $3x + 14x$

11. $17x + 18x$

12. $7p + 8p$

13. $a + 2a + 9a + 3a$

14. $x + 8x + 9x + 2x$

15. $11a + 16a$

16. $x + 3x + 13x$

17. $7x + 17x + x$

18. $y + 4y + 5y$

19. $3t + 4t + 8t + t$

20. $5t + 2t + 6t + t$

Exercise 14

Give simpler answers, **where possible** for:

1. $x + y$
2. $3x + 18x$
3. $y + a$
4. $2y + a$
5. $b + 6b + 2b + c$
6. $a + 2a + 3c$
7. $x + 3x + 7x + 2y$
8. $x + 3y$
9. $x + 11x + 3x + a + 2a$
10. $7a + 2a + 3b$
11. $5a + 3b + 3a + 5b$
12. $3x + 8x + 2x + y$
13. $a + 2b + b + 2a$
14. $x + 3y + 11y + 4x$
15. $2y + 7y + 11y + 3x$
16. $a + 2a + 3b + 5b$
17. $x + 11x + 14x - y$
18. $4x - 3x + y$
19. $13x - 8x + x + y$
20. $14a - 2a + 3c + 5c$

Look very carefully at $2a$ and a^2. **These must not be confused**.

$2a$ means $2 \times a$.
a^2 means $a \times a$.

We refer to a^2 as **'a' squared** or **'a' to the power of two**.

If $a = 3$, $a^2 = a \times a = 3 \times 3 = 9$.

Exercise 15

If $a = 2$ and $b = 4$ work answers to:

1. $2a$
2. $2b$
3. $3a$
4. $3b$
5. a^2
6. b^2
7. $2a^2$
8. $2b^2$
9. $3a^2$
10. $3b^2$
11. $\frac{1}{2}a^2$
12. $\frac{1}{2}b^2$
13. $\frac{1}{4}a^2$
14. $5a^2$
15. $5b^2$
16. $4a^2$
17. $4b^2$
18. a^2b^2
19. $2a^2b^2$
20. $3a^2b^2$

We say x **cubed** for x^3.

x^3 means $x \times x \times x$.

$2x^3$ means $2 \times x \times x \times x$.

x^4 means $x \times x \times x \times x$.

and so on.

We call x^4 **'x to the power of 4'**.

Exercise 16

If $x = 2$, $y = 4$ and $t = 6$ work answers for:

1. x^2	2. y^2	3. t^2	4. x^3
5. y^3	6. x^2y	7. y^2t	8. $3t^2$
9. $3xy^2$	10. $2y \times 3x$	11. $3t \times 2y$	12. $2x \times 3t$
13. x^4	14. $3y^2$	15. $4x^4$	16. xyt
17. $x^2y + y^2x$	18. $t^2 + x^3$	19. $y^3 + t$	20. $y^3 + t^2$

Exercise 17

If $a = 6$ work answers to:

1. $2a - 1$
2. $3a + 4$
3. $(a + 2)(8 - a)$

4. $\dfrac{2}{3}a$
5. $a + \dfrac{1}{a}$
6. $\dfrac{a}{2}$

7. $a^2 + 3a$
8. $2a^2$
9. $a(a + 3)$

10. $\dfrac{3}{a} + a^2$
11. $\dfrac{a^3}{36}$
12. $\dfrac{a}{6} + 1$

13. $\dfrac{6}{a} + a$
14. $\dfrac{a + 1}{14}$
15. $\dfrac{a^2}{72}$

16. $2a + \dfrac{4}{a}$
17. $\dfrac{18}{a} + 2$
18. $\dfrac{1}{a} + \dfrac{a}{a}$

19. $2a - 4$
20. $(a + 3)^2$

Exercise 18

If $x = 4$ and $y = 2$ work answers to:

1. $x - y$
2. $2xy$
3. $x + y$
4. $2x - y$
5. $2x + y$

6. $\dfrac{x}{y}$
7. $\dfrac{2}{y}$
8. $\dfrac{4}{x}$
9. $\dfrac{1}{x} + \dfrac{1}{y}$
10. $3x^2$

11. $4y^2$
12. $x^2 + y^2$
13. $12xy$
14. $x^2 + 2y$
15. $12x - y$

16. x^3 **17.** $\dfrac{x}{2y}$ **18.** xy^2 **19.** $(x-y)^2$ **20.** $(x-y)^3$

The **order of adding** does not matter, for example: $a + b = b + a$.

This is very easy to prove.

Suppose $a = 3$ and $b = 4$
$a + b = 3 + 4 = 7$

$b + a = 4 + 3 = 7$

So $a + b = b + a$ and we have shown that **the order of adding does not matter**.

Exercise 19

Choose your own number values for the letters and show that the order of adding does not matter in the following examples.

Show that:

1. $a + x = x + a$

2. $b + t = t + b$

3. $a + m = m + a$

4. $t + p = p + 3$

5. $b + c = c + b$

6. $p + q = q + p$

7. $a + d = d + a$

8. $a + e = e + a$

9. $a + t = t + a$

10. $m + e = e + m$

11. $a + b + t = t + b + a = a + t + b$

12. $x + y + m = y + m + x = m + x + y$

13. $e + f + g = g + e + f = f + e + g$

14. $d + c \mid a = a + c + d - c + d + a$

15. $e + d + t = t + d + e = d + e + t$

16. $p + q + r + t = t + r + q + p = q + p + t + r = r + q + p + t$

17. $e + p + t = t + p + e = p + t + e$

18. $x + m + e = e + m + x = m + x + e$

19. $a + t + d = d + a + t = t + d + a$

20. $e + f + r = r + f + e = f + r + e$

Suppose we want to show that the order of adding does not matter in the expression $2a + 3b$.
Show that $2a + 3b = 3b + 2a$.
Let $a = 4$ and $b = 2$.
$2a + 3b = 2 \times 4 + 3 \times 2 = 8 + 6 = 14$.
Remember to **multiply** first?
$3b + 2a = 3 \times 2 + 2 \times 4 = 6 + 8 = 14$ so $2a + 3b = 3b + 2a$ and therefore **the order of adding does not matter.**

Exercise 20

Choose your own number values for the letters and show that the order of adding does not matter.

Show that:

1. $2a + c = c + 2a$

2. $5p + q = q + 5p$

3. $2a + 3d = 3d + 2a$

4. $5m + n = n + 5m$

5. $3p + 5q = 5q + 3p$

6. $a + 7b = 7b + a$

7. $3c + 5a = 5a + 3c$

8. $5m + 2a = 2a + 5m$

9. $3p + 2t = 2t + 3p$

10. $5m + 3q = 3q + 5m$

11. $2t + 4q = 4q + 2t$

12. $2a + 6b = 6b + 2a$

13. $10m + 3t = 3t \times 10m$

14. $5a + 6e = 6e + 5a$

15. $3a + 2b + 8c = 8c + 2b + 3a = 2b + 8c + 3a$

16. $2m + 3n + 4q = 4q + 3n + 2m = 3n + 2m + 4q$

17. $a + 2b + 3c = 3c + a + 2b = 2b + 3c + a$

18. $m + 4n + p = p + 4n + m = 4n + p + m$

19. $6q + 8r + 2t = 2t + 8r + 6q = 8r + 6q + 2t$

20. $5a + 3b + p = p + 3b + 5a = 3b + 5a + p$

Multiplication

$2 \times 3 = 6$
Also $3 \times 2 = 6$

In multiplication, the **order** of multiplication does not matter, for example:

$2 \times 3 \times 5 \times 10 = 3 \times 5 \times 10 \times 2 = 300$.

In the same way, $a \times b = b \times a$.

Also $a \times b \times c = b \times c \times a = b \times a \times c$

Let $a = 2, b = 3$ and $c = 4$.

$a \times b \times c = 2 \times 3 \times 4 = 24$.

$b \times c \times a = 3 \times 4 \times 2 = 24$.

$b \times a \times c = 3 \times 2 \times 4 = 24$.

So **the order of multiplying does not matter.**

We usually write $3a$ and not $a3$; $4p$ and not $p4$; $3ab$ and not $a3b$. Put the number **first**.

Exercise 21

Choose number values for the letters and show that the order of multiplying does not matter in the following.

1. $ab = ba$
2. $dc = cd$
3. $abc = bca = cab$
4. $2ab = 2ba$
5. $3pq = 3qp$
6. $4cd = 4dc$
7. $5xy = 5yx$
8. $aed = dea = ade$
9. $4mn = 4nm$
10. $5tq = 5qt$
11. $adt = tda = atd$
12. $4pm = 4mp$
13. $5bc = 5cd$
14. $6rt = 6tr$
15. $adx = xda = xad$
16. $5rm = 5mr$
17. $10pt = 10tp$
18. $6ra = 6ar$
19. $\frac{1}{2}cd = \frac{1}{2}dc$
20. $\frac{1}{4}mn = \frac{1}{4}nm$

The following examples will help you with the questions which follow.

Example 1: $2x \times 3y = 6xy$

Example 2: $3a \times 2b \times c = 6abc$

Example 3: $4t \times 2p = 8tp$

Example 4: $3x \times 2x = 6x^2$

Example 5: $3x \times 3x \times 5y = 45x^2y$

In the questions which follow, **always put the number first.** We say $6c$ and never $c6$, for example:

Exercise 22

Work answers to:

1. $2x \times 3p$
2. $3x \times 2p$
3. $4a \times 8b$
4. $3b \times 2a$
5. $5a \times 2p$
6. $3a \times 2p$
7. $4b \times 2b$
8. $5y \times 2y$
9. $6m \times 2m$
10. $4t \times 2t \times 8$
11. $5p \times 6p \times 2$
12. $2a \times 6a \times 4$
13. $5a \times 2b \times c$
14. $a \times 2b \times 4d$
15. $a \times 5b \times 2c$
16. $x \times 2y \times 3t$
17. $2a \times 7b$
18. $5p \times 2q \times 6r$
19. $3p \times 4p \times 5$
20. $2x \times 3x \times 14$
21. $3a \times 5m$
22. $5p \times 7p$
23. $10t \times 2t$
24. $5a \times 2b \times 8c$
25. $5c \times 2d$
26. $3t \times 7c$
27. $2t \times 3a \times 6p$
28. $5m \times 2t \times 3c$
29. $8a \times 2b$
30. $2c \times 3m$
31. $4c \times 8d$
32. $2c \times 8d$
33. $5m \times 2r$
34. $6a \times 2m$
35. $2t \times 16t$
36. $5c \times 8e$
37. $2m \times 18r$
38. $3t \times 16t$
39. $5p \times 2a$
40. $2a \times 2b \times 2c$

Exercise 23(a)

If $a = 2, b = 1$ and $c = 3$ work number answers for:

1. $2a$
2. $3b$
3. $5c$
4. $2ab$
5. $3bc$
6. $7ac$
7. a^2
8. b^2
9. c^2
10. $2abc$
11. b^2c
12. c^2b
13. $3a^2b$
14. $2b^2a$
15. c^2ab
16. a^2bc
17. $4a \times 2b$
18. $2c \times 3a$
19. $2a \times 3b \times 4c$
20. $a \times 2b \times 5c$
21. $5a \times 11b$

22. $a \times 3c \times 4b$ **23.** $a^2b^2c^2$ **24.** $6ab^2$

25. $5a^2b$ **26.** $3c^2a$ **27.** $12b^2c$

28. $4a^2c^2$ **29.** a^2b^2 **30.** a^3b^2

31. c^2a^3 **32.** c^3b^2 **33.** $2c^3b$

34. $11ab$ **35.** b^3c **36.** $\dfrac{abc}{6}$

37. $\frac{1}{2}a^3$ **38.** $2a \times 5b \times c$ **39.** $3b^3$

40. $3c^2a^3$

Exercise 23(b)

Repeat exercise 23(a) using $a = 1$, $b = 3$ and $c = 2$.

Division

Example 1: $\dfrac{\overset{2}{\cancel{10}}ab}{\cancel{5}_1} = 2ab$

Example 2: $\dfrac{3p^2q}{6q^2} = \dfrac{\overset{1}{\cancel{3}} \times p \times p \times \cancel{q}^{\,1}}{\cancel{6} \times q \times \cancel{q}_1 \,\underset{2}{}} = \dfrac{p^2}{2q}$

Example 3: $\dfrac{2a^2b}{ab} = \dfrac{2 \times \cancel{a}^{\,1} \times a \times \cancel{b}^{\,1}}{_1\cancel{a} \times \cancel{b}_1} = 2a$

Example 4: $\dfrac{20x^2}{5x} = \dfrac{\overset{4}{\cancel{20}} \times x \times \cancel{x}^{\,1}}{\cancel{5} \times \cancel{x}_{1}\,{}_1} = 4x$

Exercise 24

Work answers to the following:

1. $\dfrac{10ab}{2b}$

2. $\dfrac{15xy}{5x}$

3. $\dfrac{2xy^2}{y}$

4. $\dfrac{2m^2n}{mn}$

5. $\dfrac{3p^2}{p}$

6. $\dfrac{5abc}{bc}$

7. $\dfrac{4pq^2}{2q}$

8. $\dfrac{3ab}{6}$

9. $\dfrac{5m^2t}{mt}$

10. $\dfrac{4ab}{2b}$

11. $\dfrac{3a^2}{6a}$

12. $\dfrac{5pq}{p}$

13. $\dfrac{5ax}{x^2}$

14. $\dfrac{3ab}{b^2}$

15. $\dfrac{5ma}{a^2}$

16. $\dfrac{def}{def}$

17. $\dfrac{10a^2b}{5b^2}$

18. $\dfrac{3p^3q}{p^2}$

19. $\dfrac{14a^2}{7a}$

20. $\dfrac{16a^2x}{8x}$

21. $\dfrac{40a^2b}{20a}$

22. $\dfrac{8c^2d}{4d}$

23. $\dfrac{5de^2}{e}$

24. $\dfrac{x^2y}{y^2x}$

25. $\dfrac{15p^3}{5p}$

26. $\dfrac{30a^2b}{15}$

27. $\dfrac{2x^2y}{4y}$

28. $\dfrac{5abc}{bc}$

29. $\dfrac{12a^2m}{6a}$

30. $\dfrac{3p^3q}{6q}$

31. $\dfrac{5d^2}{10d}$

32. $\dfrac{6de}{d^2}$

33. $\dfrac{8at}{4a}$

34. $\dfrac{6abt}{2a^2}$

35. $\dfrac{xyt}{2x}$

36. $\dfrac{10p}{5}$

37. $\dfrac{15m^2t}{3mt}$

38. $\dfrac{4p^2t}{2t}$

39. $\dfrac{25a^3}{5a^2}$

40. $\dfrac{4x^2t}{2t^2x}$

Formulae

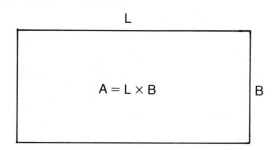

$$A = L \times B$$

Here we have a rectangle, a shape which we all recognise. The **area** is given by $A = L \times B$. The **perimeter** is given by $P = 2L + 2B$.

We say that $A = L \times B$ and $P = 2L + 2B$ are **formulae.** A formula is like a recipe. If we use the formula we get the result by using the **data** according to a certain rule.

Example 1: A rectangle has a length of 3 metres and a breadth of 2 metres. Using $A = L \times B$ and $P = 2L + 2B$ work out the **area in square metres** and the **perimeter in metres.**

        $L = 3$ and $B = 2$ are the **data.**
        $A = L \times B$
        gives us $A = 3 \times 2 = 6$
        so the area is 6 metres²
        $P = 2L + 2B$
        gives us $P = 2 \times 3 + 2 \times 2 = 6 + 4 = 10$.

So the perimeter is 10 metres.

Use the formulae $A = L \times B$ and $P = 2L + 2B$ to find A and P for each of the following.

The unit of length is the centimetre throughout the exercise.

Exercise 25

1. $L = 4, B = 2$ 2. $L = 6, B = 4$ 3. $L = 1, B = 1$

4. $L = 4, B = 3$ 5. $L = 10, B = 7$ 6. $L = 10, B = 9$

7. $L = 20, B = 8$ 8. $L = 20, B = 10$ 9. $L = 14, B = 13$

10. $L = 14, B = 7$ 11. $L = 6, B = 2\frac{1}{2}$ 12. $L = 5, B = 3\frac{1}{2}$

13. $L = 2, B = 1\frac{1}{2}$ 14. $L = 7, B = 5\frac{1}{2}$ 15. $L = 30, B = 20$

16. $L = 24, B = 20$ 17. $L = 18, B = 10$ 18. $L = 16\frac{1}{2}, B = 15$

19. $L = 14, B = 10$ 20. $L = 14, B = 12$ 21. $L = 8, B = 6\frac{1}{2}$

22. $L = 8, B = 7\frac{1}{2}$ 23. $L = 9, B = 1\frac{1}{2}$ 24. $L = 9, B = 6$

25. $L = 9, B = 4\frac{1}{2}$ 26. $L = 10, B = 7\frac{1}{2}$ 27. $L = 10, B = 2\frac{1}{2}$

28. $L = 20, B = 15$ **29.** $L = 10, B = 3\frac{1}{2}$ **30.** $L = 2, B = 1$

31. $L = 17, B = 15$ **32.** $L = 16, B = 10\frac{1}{2}$ **33.** $L = 4, B = 3\frac{1}{2}$

34. $L = 16, B = 12\frac{1}{2}$ **35.** $L = 24, B = 18$ **36.** $L = 32, B = 20$

37. $L = 32, B = 16$ **38.** $L = 5\frac{1}{2}, B = 2\frac{1}{2}$ **39.** $L = 6\frac{1}{2}, B = 2$

40. $L = 7\frac{1}{2}, B = 3$

Exercise 26

The formula $D = S \times T$ connects distance, speed and time. The time is in hours and the speed is in km/h in each case. The distances are all in kilometres.

Using the formula $D = S \times T$, find D in each of the following.

1. $S = 30, T = 1$ **2.** $S = 30, T = 4$ **3.** $S = 50, T = 1$

4. $S = 60, T = 4$ **5.** $S = 40, T = 2$ **6.** $S = 50, T = 2$

7. $S = 80, T = 1$ **8.** $S = 80, T = 2$ **9.** $S = 100, T = 2$

10. $S = 100, T = 3$ **11.** $S = 70, T = 2$ **12.** $S = 70, T = 4$

13. $S = 50, T = 2\frac{1}{2}$ **14.** $S = 40, T = 3\frac{1}{2}$ **15.** $S = 80, T = 10$

16. $S = 80, T = 8\frac{1}{2}$ **17.** $S = 30, T = 6$ **18.** $S = 30, T = 10$

19. $S = 60, T = \frac{1}{2}$ **20.** $S = 100, T = \frac{1}{4}$ **21.** $S = 90, T = 4$

22. $S = 54, T = \frac{1}{2}$ **23.** $S = 35, T = 2$ **24.** $S = 85, T = 3$

25. $S = 110, T = 2$ **26.** $S = 64. T = 5$ **27.** $S = 72, T = \frac{3}{4}$

28. $S = 80, T = 1\frac{1}{2}$ **29.** $S = 90, T = 2\frac{1}{2}$ **30.** $S = 48, T = 6$

31. $S = 62, T = 8$ **32.** $S = 25, T = 5$ **33.** $S = 46, T = \frac{1}{2}$

34. $S = 100, T = \frac{1}{4}$ **35.** $S = 84, T = \frac{1}{4}$ **36.** $S = 96, T = 2$

37. $S = 65, T = \frac{1}{5}$ **38.** $S = 45, T = \frac{1}{5}$ **39.** $S = 28, T = 4$

40. $S = 68, T = 2$

Exercise 27

The formula to be used now is $t = a + nd$.

In each case you are given a, n and d. Work out t.

1. $a = 2, n = 1, d = 3$ **2.** $a = 1, n = 3, d = 2$ **3.** $a = 4, n = 3, d = 2$

4. $a = 5, n = 4, d = 1$ **5.** $a = 6, n = 5, d = 3$ **6.** $a = 4, n = 1, d = 6$

7. $a = 8, n = 6, d = 1$ **8.** $a = 14, n = 3, d = 2$ **9.** $a = 6, n = 8, d = 10$

10. $a = 10, n = 1, d = 5$ 11. $a = 6, n = 10, d = 6$ 12. $a = 2, n = 4, d = 6$

13. $a = 5, n = 5, d = 2$ 14. $a = 7, n = 4, d = 1$ 15. $a = 4, d = 14, n = 2$

16. $a = 1, n = 20, d = 2$ 17. $a = 8, d = 14, n = 16$ 18. $a = 3, n = 5, d = 7$

19. $a = 4, d = 4, n = 8$ 20. $a = 2, n = 2, d = 17$ 21. $a = 1, n = 2, d = 4$

22. $a = 5, n = 6, d = 2$ 23. $a = 2, n = 10, d = 3$ 24. $a = 15, n = 4, d = 1$

25. $a = 16, n = 8, d = 2$ 26. $a = 1, n = 10, d = 20$ 27. $a = 4, n = 16, d = 5$

28. $a = 2, n = 14, d = 10$ 29. $a = 1, n = 18, d = 4$ 30. $a = 18, n = 4, d = 2$

31. $a = 24, n = 1, d = 3$ 32. $a = 8, n = 2, d = 12$ 33. $a = 6, n = 14, d = 2$

34. $a = 1, n = 10, d = \frac{1}{2}$ 35. $a = 4, n = 20, d = \frac{1}{4}$ 36. $a = 6, n = 10, d = 200$

37. $a = 4, n = 8, d = 16$ 38. $a = 1, n = 1, d = 1000$ 39. $a = 1, n = 4, d = 7$

40. $a = 2, n = 100, d = 10$

Exercise 28

Now we will use a formula which states that $T = \dfrac{P}{Q + R}$.

In each case work out T if you are given P, Q and R.

Leave your answer, in each case, as a fraction in its lowest terms, for example:

if $P = 50, Q = 10$ and $R = 20, T = \dfrac{P}{Q + R} = \dfrac{50}{10 + 20} = \dfrac{50}{30} = \dfrac{5}{3}$.

1. $P = 2, Q = 8, R = 7$ 2. $P = 3, Q = 1, R = 2$

3. $P = 5, Q = 2, R = 3$ 4. $P = 9, Q = 10, R = 7$

5. $P = 20, Q = 18, R = 2$ 6. $P = 40, Q = 4, R = 16$

7. $P = 50, Q = 10, R = 20$ 8. $P = 18, Q = 14, R = 14$

9. $P = 25, Q = 16, R = 8$ 10. $P = 200, Q = 50, R = 40$

11. $P = 100, Q = 70, R = 80$ 12. $P = 6, Q = 4, R = 5$

13. $P = 80, Q = 50, R = 20$ 14. $P = 40, Q = 40, R = 10$

15. $P = 10, Q = 20, R = 50$ 16. $P = 18, Q = 2, R = 7$

17. $P = 10, Q = 2, R = 3$ 18. $P = 50, Q = 10, R = 5$

19. $P = 35, Q = 3, R = 4$ 20. $P = 65, Q = 7, R = 6$

21. $P = 10, Q = 7, R = 1$ 22. $P = 1, Q = 2, R = 10$

23. $P = 6, Q = 12, R = 6$ 24. $P = 100, Q = 10, R = 70$

25. $P = 40, Q = 2, R = 3$ 26. $P = 16, Q = 2, R = 6$

27. $P = 20. Q = 1. R = 9$ 28. $P = 400. Q = 60. R = 20$

29. $P = 60. Q = 10. R = 10$ 30. $P = 30. Q = 7. R = 8$

31. $P = 50. Q = 13. R = 12$ 32. $P = 90. Q = 4. R = 5$

33. $P = 30. Q = 15. R = 5$ 34. $P = 75. Q = 15. R = 10$

35. $P = 20. Q = 1. R = 0$ 36. $P = 0. Q = 1. R = 1$

37. $P = 60. Q = 30. R = 30$ 38. $P = 150. Q = 25. R = 25$

39. $P = 120. Q = 30. R = 30$ 40. $P = 8. Q = 1. R = 7$

Use of Brackets

When we write $2(x + y)$ this is the same as $2x + 2y$.

When we write $5(2a + 3c)$ this is the same as $10a + 15c$.

When we write $7(2x + 8y)$ this is the same as $14x + 56y$.

When we write $2(4x - y)$ this is the same as $8x - 2y$.

Exercise 29

Write without brackets:

1. $3(2a + c)$
2. $(2b + c)$
3. $15(2a + 3c)$
4. $7(x + 2y)$
5. $8(3x - y)$
6. $6(2a + 5c)$
7. $4(2x - 8y)$
8. $5(2p + q)$
9. $4(a + 9d)$
10. $3(6x - 2y)$
11. $4(5x + y)$
12. $3(3x + 20y)$
13. $4(x - 6y)$
14. $8(5x - 3y)$
15. $2(2x + 5a)$
16. $8(p + 7q)$
17. $6(2p - 5q)$
18. $10(3x + y)$
19. $10(4x - 3y)$
20. $5(7x - 6y)$
21. $5(2p + q)$
22. $7(p - q)$
23. $10(2p + 1)$
24. $5(3p - q)$
25. $6(a + 3b)$
26. $7(2a - b)$
27. $6(2a - 3b)$
28. $7(2a + b)$
29. $6(a - 3b)$
30. $3(x + 2y)$
31. $5(2y - t)$
32. $6(a + 3t)$
33. $2(2a - b)$
34. $5(3m + 2n)$
35. $6(m - 3n)$
36. $8(2a + 11b)$
37. $2(4a - 3c)$
38. $\frac{1}{2}(6a + 8c)$
39. $\frac{1}{4}(4b + 8t)$
40. $10(2b + 3a)$
41. $5(2m + 3t)$
42. $3(6a - b)$
43. $2(5m + 11)$
44. $2(3t + 6a)$
45. $3(2 + 5a)$
46. $4(7a - c)$
47. $2(5a + 6c)$
48. $3(5a - r)$
49. $5(6a + 1)$
50. $2(7a - 3)$
51. $3(6t + 5)$
52. $4(x + 8y)$
53. $3(2x - 6y)$
54. $4(x - 18)$
55. $\frac{1}{2}(6x + 8y)$
56. $10(3a - d)$
57. $12(3a + d)$
58. $20(\frac{1}{2} + a)$
59. $40(\frac{1}{4} + t)$
60. $2(m + 10t)$

A **simple** equation has only **one** unknown. What is an **unknown**?

Look at this: $3 + 4 = 7$

Every position here is filled by a **quantity we know.**

Look at this: $x + 7 = 15$

One position here is filled by an **unknown** quantity. We say that $x + 7 = 15$ is an example of a simple equation. The unknown quantity is x.

To **solve** the equation we need to know the value of the **unknown quantity**. When $x + 7 = 15$, x will equal 8. Do you agree?

If we look at the equation $p + 7 = 18$ we can say it is a **simple** equation because **there is only one unknown**.

If $p + 7 = 18$, p must equal 11.

We say that $p = 11$ is the **solution** of the equation.

Exercise 30

Solve the simple equations:

1. $x + 3 = 4$	2. $p + 6 = 7$	3. $x + 9 = 12$
4. $t + 11 = 13$	5. $y + 3 = 8$	6. $m + 2 = 19$
7. $a + 11 = 14$	8. $c + 2 = 3$	9. $x + 1 = 14$
10. $x + 7 = 19$	11. $x + 17 = 25$	12. $x + 9 = 19$
13. $x + 2 = 29$	14. $x + 10 = 29$	15. $x + 7 = 14$
16. $a + 9 = 26$	17. $x + 4 = 16$	18. $x + 13 = 17$
19. $y + 2 = 23$	20. $m + 1 = 46$	21. $t + 11 = 17$
22. $t + 6 = 16$	23. $x + 9 = 40$	24. $x + 8 = 41$
25. $a + 7 = 100$	26. $a + 11 = 94$	27. $t + 3 = 37$
28. $t + 12 = 84$	29. $a + 1 = 106$	30. $x + 3 = 104$
31. $n - 4 = 16$	32. $e + 5 = 11$	33. $d + 1 = 27$
34. $p - 6 = 13$	35. $a + 2 = 11$	36. $m - 5 = 15$
37. $m + 6 = 47$	38. $t - 3 = 17$	39. $t + 5 = 98$
40. $t - 3 = 27$		

Exercise 31

Find a solution for each of the following equations.

1. $a - 1 = 3$
2. $c - 3 = 4$
3. $t - 5 = 6$
4. $m - 6 = 17$
5. $a - 1 = 13$
6. $x - 7 = 15$
7. $a - 4 = 11$
8. $x - 2 = 17$
9. $y - 3 = 9$
10. $a - 6 = 16$
11. $t - 3 = 7$
12. $c - 1 = 100$
13. $t - 3 = 14$
14. $p - 5 = 16$
15. $q - 3 = 39$
16. $a - 2 = 17$
17. $c - 14 = 14$
18. $a - 3 = 3$
19. $m - 20 = 41$
20. $m - 16 = 41$
21. $e - 7 = 11$
22. $e + 2 = 19$
23. $f + 3 = 4$
24. $f - 2 = 6$
25. $g + 1 = 7$
26. $g - 2 = 9$
27. $k + 1 = 10$
28. $k - 3 = 14$
29. $k + 6 = 7$
30. $k + 11 = 19$
31. $h + 2 = 23$
32. $h - 6 = 14$
33. $h - 1 = 2$
34. $h + 1 = 3$
35. $q + 10 = 20$
36. $q - 1 = 40$
37. $q + 5 = 31$
38. $q - 6 = 32$
39. $q + 10 = 104$
40. $q - 10 = 60$

Exercise 32

Find a solution for:

1. $x + 4 = 27$
2. $x + 6 = 79$
3. $x + 4 = 10$
4. $p - 7 = 106$
5. $a + 1 = 12$
6. $b + 3 = 18$
7. $w + 2 = 24$
8. $x - 3 = 14$
9. $x + 7 = 81$
10. $x - 1 = 19$
11. $x - 3 = 102$
12. $x - 2 = 71$
13. $x - 1 = 23$
14. $x + 1 = 300$
15. $x - 2 = 109$
16. $x - 1 = 40$
17. $a - 7 = 47$
18. $a - 111 = 2$
19. $t - 31 = 1000$
20. $c - 32 = 46$
21. $v + 11 = 12$
22. $v - 6 = 10$
23. $v + 1 = 103$
24. $v - 6 = 8$
25. $y + 5 = 73$
26. $y - 5 = 12$
27. $y - 3 = 49$
28. $y + 3 = 12$
29. $y + 10 = 49$
30. $y - 10 = 2$
31. $e + 11 = 103$
32. $e - 11 = 24$
33. $f + 17 = 49$

34. $f - 17 = 18$	**35.** $g + 1 = 100$	**36.** $g - 6 = 17$
37. $t + 11 = 41$	**38.** $t - 13 = 17$	**39.** $t - 16 = 1$
40. $t + 10 = 17$		

So far you have been able to 'read off' your answer just by looking at the equation.

Now we will move on to a slightly more difficult type.

Suppose $2x + 5 = 19$.

What, with 5 added, gives 19?

The answer, of course, is 14.

So $2x = 14$.

2 times what equals 14? $2 \times 7 = 14$.

So x must be 7 in this case.

Test: $2x + 5 = 19$
$3x + 5 = 2 \times 7 + 5 = 14 + 5 = 19$: if $x = 7$.

Example: Solve for x, $3x + 1 = 16$
What, with one added, gives 16?
The answer, of course, is 15.
So $3x = 15$.
3 times what equals 15?
$3 \times 5 = 15$
So $x = 5$ in this case.

Test: $3x + 1 = 3 \times 5 + 1 = 15 + 1 = 16$: if $x = 5$.

Exercise 33

Work out the value of the unknown in each of the following.

1. $3a + 1 = 7$	**2.** $5t + 4 = 19$	**3.** $2x + 5 = 11$
4. $3x + 5 = 8$	**5.** $5x + 1 = 16$	**6.** $7t + 13 = 27$
7. $2x + 8 = 12$	**8.** $5x + 1 = 21$	**9.** $3a + 13 = 25$
10. $2t + 9 = 11$	**11.** $7c + 1 = 36$	**12.** $10a + 1 = 11$
13. $9x + 8 = 35$	**14.** $17t + 2 = 19$	**15.** $14t + 3 = 31$
16. $4x + 1 = 29$	**17.** $3x + 11 = 41$	**18.** $12a + 5 = 17$
19. $x + 13 = 14$	**20.** $2x + 11 = 11$	**21.** $13a + 9 = 22$

22. $3a + 17 = 17$	23. $2x + 32 = 35$	24. $13a + 1 = 40$
25. $8x + 7 = 23$	26. $15a + 1 = 31$	27. $19x + 3 = 41$
28. $2z + 14 = 14$	29. $3t + 1 = 40$	30. $3x + 8 = 32$
31. $2e + 9 = 14$	32. $5e - 1 = 19$	33. $3g + 11 = 14$
34. $7p + 11 = 25$	35. $3t - 1 = 101$	36. $7x - 5 = 30$
37. $11t - 1 = 21$	38. $2a - 6 = 7$	39. $2a - 3 = 4$
40. $2a - 4 = 5$		

Exercise 34

Work out the value of the unknown in each of the following.

1. $5a - 1 = 99$	2. $20x - 3 = 17$	3. $13t - 2 = 24$
4. $4x + 7 = 11$	5. $30m - 1 = 29$	6. $2a - 1 = 19$
7. $16x - 7 = 9$	8. $10x - 11 = 19$	9. $9a - 1 = 8$
10. $8x - 3 = 13$	11. $5m - 3 = 2$	12. $6x - 1 = 23$
13. $3a - 5 = 10$	14. $11x - 24 = 31$	15. $10x - 1 = 19$
16. $3a - 7 = 2$	17. $14x - 1 = 13$	18. $12a - 11 = 1$
19. $20x - 3 = 77$	20. $2p - 1 = 5$	21. $2x - 1 = 3$
22. $7a - 3 = 4$	23. $12x - 7 = 17$	24. $4a - 11 = 21$
25. $5x - 18 = 2$	26. $2z - 5 = 19$	27. $11a - 1 = 10$
28. $4x - 29 = 3$	29. $20x - 3 = 37$	30. $5x - 1 = 24$

Geometry

Straight Lines

Exercise 1

Draw straight lines which measure:

1. 4 cm	**2.** 9 cm	**3.** 10 cm	**4.** 8 cm	**5.** 1 cm
6. 3 cm	**7.** 2 cm	**8.** 4 cm	**9.** 6 cm	**10.** 5 cm

Draw straight lines which measure:

1. 4·6 cm	**2.** 9·5 cm	**3.** 10·2 cm	**4.** 8·9 cm	**5.** 1·8 cm
6. 3·4 cm	**7.** 2·3 cm	**8.** 4·6 cm	**9.** 6·4 cm	**10.** 5·7 cm

Draw straight lines which measure:

1. 48 mm	**2.** 96 mm	**3.** 102 mm	**4.** 68 mm	**5.** 83 mm
6. 142 mm	**7.** 38 mm	**8.** 92 mm	**9.** 110 mm	**10.** 47 mm

Draw straight lines which measure:

1. 3·2 inches	**2.** 4·8 inches	**3.** 5·6 inches	**4.** 7·2 inches
5. 3·6 inches	**6.** 4·9 inches	**7.** 4·7 inches	**8.** 3·9 inches
9. 6·2 inches	**10.** 5·3 inches		

Exercise 2

In this exercise, look at each line and **estimate** its length, then check by measuring the line with your ruler. Give your length first in centimetres and then in inches. You will see that each **line** has a name. We put a capital letter at each end and call our lines AB or CD or GK etc.

1.
A   C
———————————————————————————————

2.
K   L
———————————————————————————————————

3.
M   N
——————————

4. C ———————————————————— D

5. A ———————————————————— K

6. L ———————————————————— M

7. M ———————————————————— N

8. C ———————————————————— D

9. A ———————————————————— R

10. N ———————————————————— L

When two straight lines cross they cross at a point.

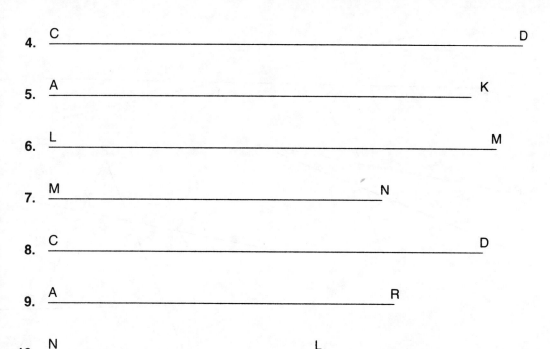

The lines AB and CD cross at point T.
We now have **line segments** in the diagram above.
CT, TD, AT, TB, AB and CD are all the lines or line segments in the diagram above.

Exercise 3

Name all the lines and line segments you can find in the following diagrams.

1.

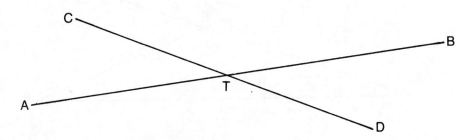

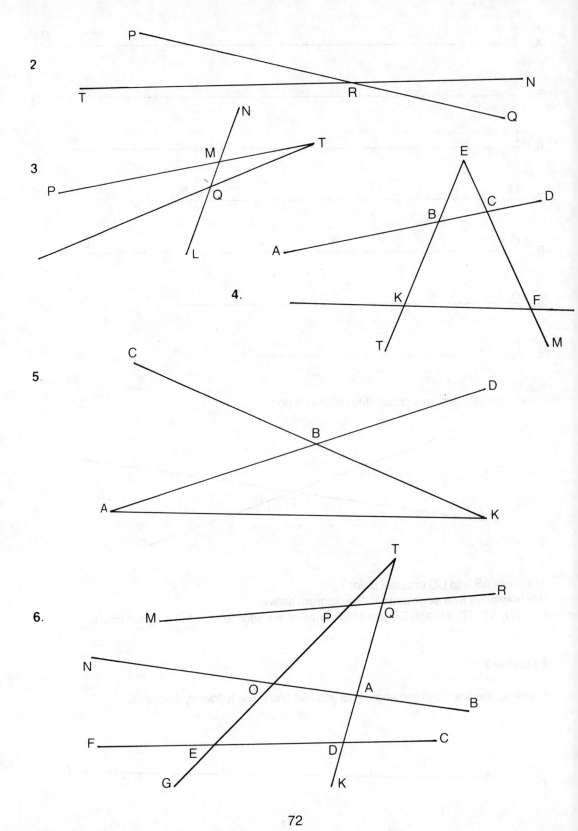

2.

3.

4.

5.

6.

When you have named as many lines and line segments as you can find, take your ruler and measure each line in centimetres. Make a table for each of the questions **1.** to **6.** to show your results.

Exercise 4

In each case, choose letters to name your lines.

1. Draw a line 2·6 cm long.
    Draw a second line **twice** as long.

2. Draw a line 10·8 cm long.
    Draw a second line **half** as long.

3. Draw a line 12·6 cm long.
    Draw a second line **one-third** as long.

4. Draw a line 12 cm long.
    Draw a second line **two-thirds** as long.

5. Draw a line 5·2 inches long.
    Draw a second line 2·6 inches long.
    What **fraction** of the first line is the second?

6. Draw a line 3·2 inches long.
    Draw a second line one-quarter as long.

7. Draw a line 13 cm long.
    Draw a second line one-tenth as long.

8. Draw a line 5 cm long.
    Draw a second line $2\frac{1}{2}$ times as long.

9. Draw a line 4 inches long.
    Draw a second line three-quarters as long.

10. Draw a line 6 cm long.
    Draw a second line one and a half times as long.

Be sure you know what a vertical line and a horizontal line look like.

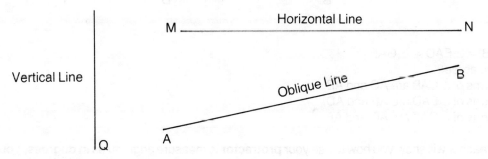

AB which is neither vertical nor horizontal is an **oblique** line.

Exercise 5

When two straight lines meet at a point, **an angle** is formed.

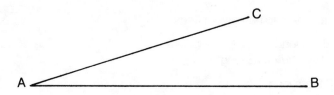

The straight lines AB and AC meet at point A.
AB and AC are called the **arms** of angle CAB.
A is the vertex.
The angle may be named as CAB or as BAC. The vertex A must be the **centre letter**.
For angle use ∠
So angle PQR is written as ∠ PQR.
The lengths of the arms of the angle **do not affect the size** of the actual angle. Look at the diagram:

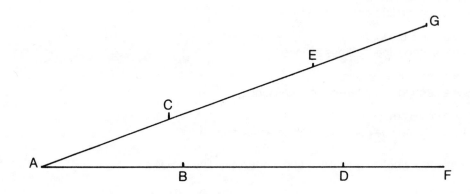

∠ CAB = ∠ EAD = ∠ GAF.
Do you agree?
The arms of ∠ CAB are AC and AB.
The arms of ∠ EAD are AE and AD.
The arms of ∠ GAF are AG and AF.

Your teacher will show you how to use your **protractor** to measure angle sizes **in degrees**. You should be able to **estimate** the size of an angle by looking at it. It is helpful to remember that **a right-angle** is 90 degrees.

We all know what a right-angle looks like, don't we?

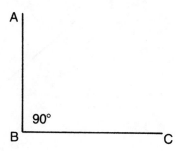

∠ ABC is a right-angle.

Sometimes we mark the angle in this way:

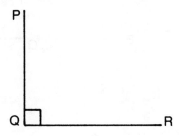

By looking at an angle of 90 degrees, can you 'picture' an angle of 45°?

Can you 'picture' an angle of 30°?
Can you 'picture' an angle of 120°?

Exercise 6

Estimate the size of each of the following angles. Check to see how 'good' your estimates are by using your protractor.

1.

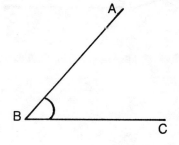

2.

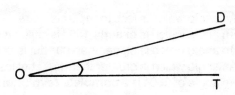

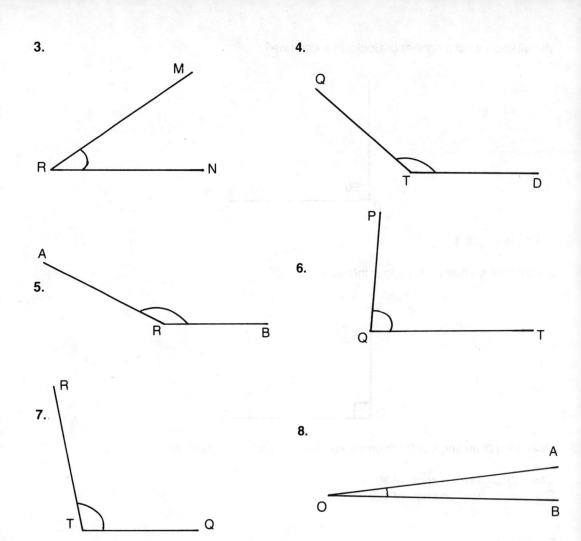

3.

4.

5.

6.

7.

8.

9.

10.

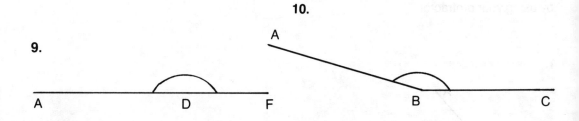

An angle which is less than 90° is **acute**.
An angle which is **exactly 180°** is called **a straight angle**.
An angle which is bigger than 90° but less than 180° is **obtuse**.
An angle which is greater than 180° but less than 360° is **reflex**.
An angle of 360° is **a complete revolution**.

In the following exercise, estimate the size of each angle, then **measure** each angle and say whether it is acute, obtuse, right, straight or reflex.

Exercise 7

1.

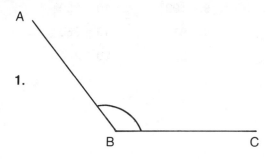

2.

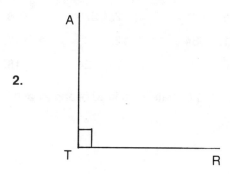

3.

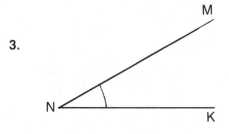

4.

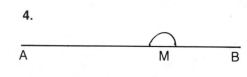

5.

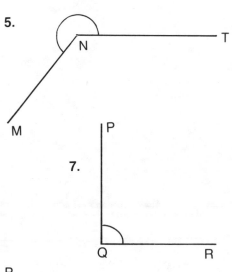

6.

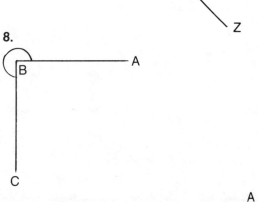

7.

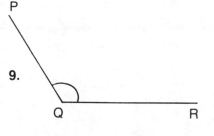

8.

9.

10.

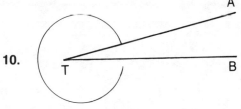

Exercise 8

Draw angles of:

1.	20°	**2.**	170°	**3.**	45°	**4.**	62°	**5.**	112°
6.	90°	**7.**	270°	**8.**	180°	**9.**	192°	**10.**	300°
11.	334°	**12.**	30°	**13.**	69°	**14.**	91°	**15.**	360°
16.	19°	**17.**	324°	**18.**	294°	**19.**	286°	**20.**	186°

In 1 to 20, state what **kind** of angle each is.

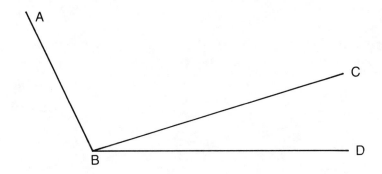

∠ ABC and ∠ CBD are **adjacent** angles. They **share** an arm.
The **shared** arm is BC. In the diagram above, if ∠ ABD = 110° and ∠ CBD = 38°, do you agree
that ∠ ABC should measure 72°?

Exercise 9

In the following examples, make a sketch of the angles before you give your answer to each
question. Then make an accurate drawing to illustrate your answers.

1. ∠ ABD = 120°
∠ ABC and ∠ CBD are adjacent angles.
The shared arm is BC and ∠ CBD = 35°
What size is ∠ ABC?

2. ∠ PQT is 98°.
∠ TQR and ∠ PQT are adjacent angles.
The shared arm is QT.
∠ PQR = 130°.
What size is ∠ TQR?

3. ∠ ABC is 240°.
∠ DBC and ∠ ABD are adjacent angles.
∠ CBD is a right angle.
BD is the shared arm.
What size is ∠ ABD?

4. ∠ BOD = 260°.
∠ BOR and ∠ ROD are adjacent angles.
OR is the shared arm.
∠ DOR = 82°.
What size is ∠ BOR?

5.   ∠ABC = 86°
   ∠ABD and ∠DBC are adjacent angles.
   BD is the shared arm.
   ∠ABD is 25°.
   Find ∠DBC.

When two straight lines cross each other at a point, four angles are formed around that point.
Look at the diagram below.

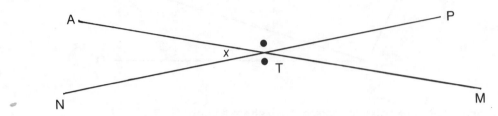

Two pairs of **vertically opposite angles** are formed around the point T.

These are (**1**) ∠ATP and ∠NTM; and (**2**) ∠ATN and ∠PTM.

Vertically opposite angles equal each other.

You will see from the figure that ∠ATP and ∠PTM are **adjacent.**
The arms TA and TM form **a straight line**.

Measure ∠ATP and ∠PTM **carefully**.
Add your results.
You should get 180°

You will see from the figure that ∠ATP and ∠ATN are **adjacent.**
The arms TP and TN form **a straight line**.
Measure ∠ATP and ∠ATN **carefully**.
Add your results.
You should get 180°
Repeat, using (**1**) ∠ATN and ∠NTM; and (**2**) ∠NTM and ∠MTP.

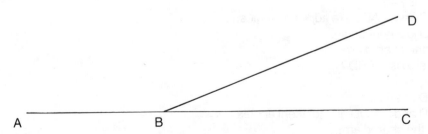

We say that when one straight line meets another straight line, two adjacent angles are formed which add to give 180°.

Exercise 10

In each of the following examples, find the sizes of **the unmarked angles. The diagrams are not drawn to scale. Name** each angle as you write down its size.

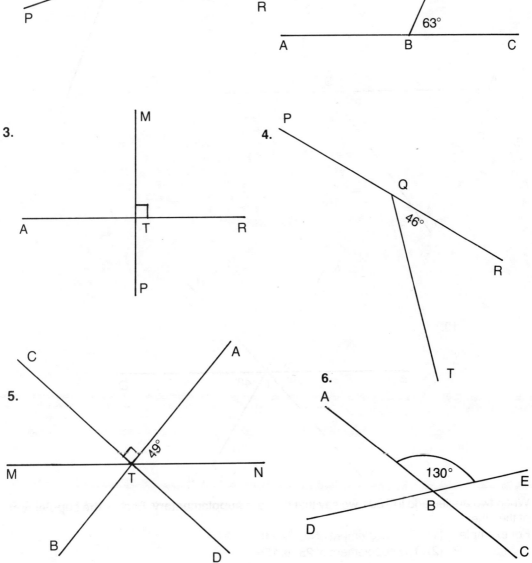

1.

2.

3.

4.

5.

6.

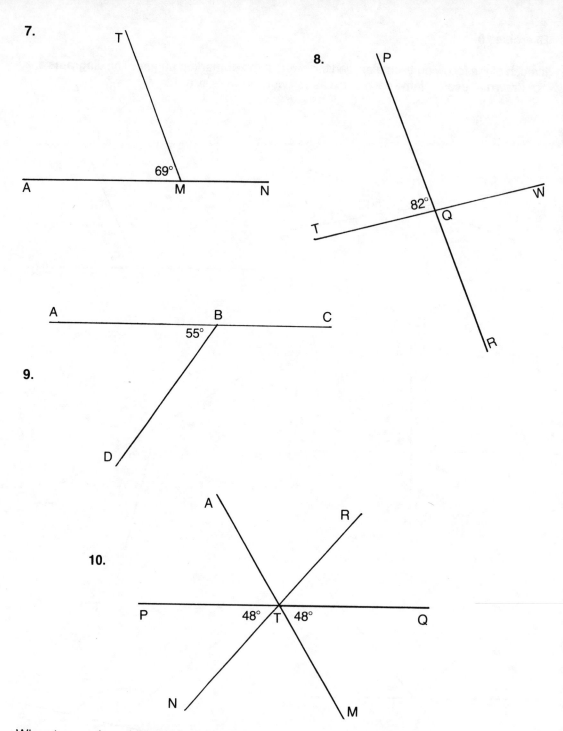

7.

8.

9.

10.

When two angles add to 180°, we say that they are **supplementary**. Each is **the supplement** of the other.

For example: **(1)** The supplement of 40° is 140°

(2) The supplement of 25° is 155°.

Exercise 11

Write down the supplement of each of the following angles.

1. 23°	**2.** 121°	**3.** 90°	**4.** 79°	**5.** 63°
6. 138°	**7.** 179°	**8.** 39°	**9.** 47°	**10.** 58°
11. 69°	**12.** 72°	**13.** 88°	**14.** 93°	**15.** 101°
16. 119°	**17.** 128°	**18.** 131°	**19.** 142°	**20.** 168°

When two angles add to 90° we say they are **complementary**.
Each is **the complement** of the other.

For example: **(1)** The complement of 20° is 70°
           (2) The complement of 40° is 50°.

Exercise 12

Write down the complement of each of the following angles.

1. 2°	**2.** 82°	**3.** 14°	**4.** 71°	**5.** 29°
6. 63°	**7.** 31°	**8.** 58°	**9.** 47°	**10.** 8°
11. 89°	**12.** 17°	**13.** 78°	**14.** 27°	**15.** 62°
16. 33°	**17.** 51°	**18.** 42°	**19.** 1°	**20.** 45°

Exercise 13

In the following figures, find the sizes of the unmarked angles. The figures are not drawn **accurately**.

1.

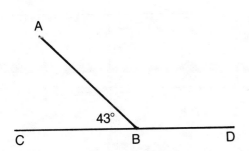

2.

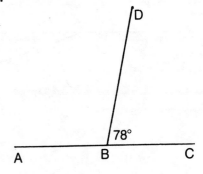

3.

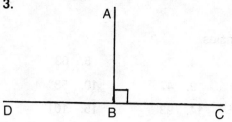

4.

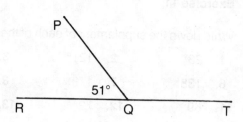

5.

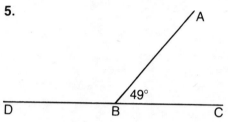

6.

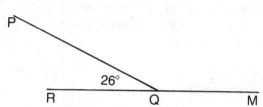

7.

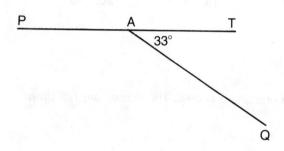

8.

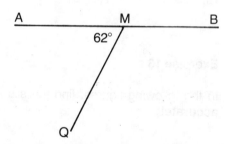

9.

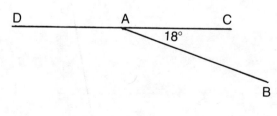

10.

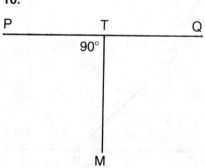

84

Exercise 14

Look at Figure 1 below and measure ∠ABF; ∠FBE; ∠EBD; ∠DBC; and ∠CBA. Add your results. You should get approximately 360°

Figure 1

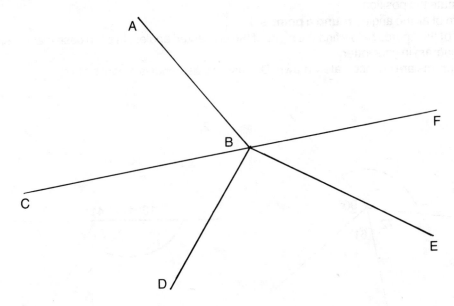

Look at Figure 2, below

Figure 2

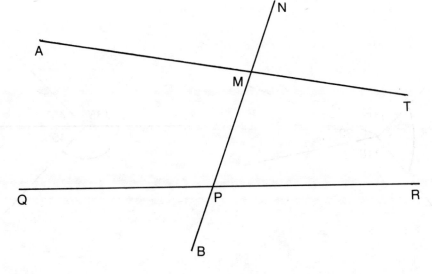

Measure ∠ NMA; ∠ NMT; ∠ AMP; and ∠ TMP.
Add your results. You should get approximately 360°.
Measure ∠ MPQ; ∠ MPR; ∠ QPB; and ∠ RPB.
Add your results. You should get approximately 360°.
In each of the above cases you will not get 360° **exactly**. This happens because it is very difficult to measure **exactly**.
Let us state the position.
The sum of all the angles **round a point** is 360°.
In each of the figures below find the sizes of the unmarked angles. In each case make a sketch of the diagram in your jotter.
The diagrams are not accurately drawn. Discuss in class what is meant by this.

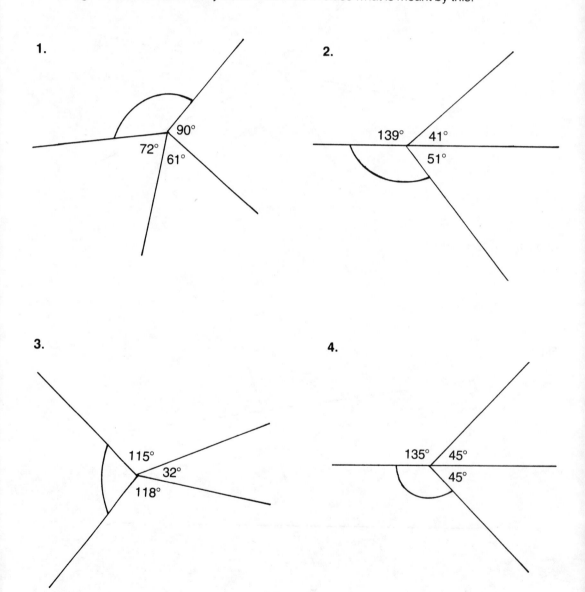

1.

90°
72°
61°

2.

139°   41°
51°

3.

115°
32°
118°

4.

135°   45°
45°

86

5.

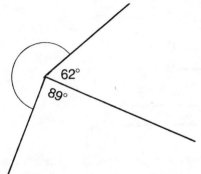

6.

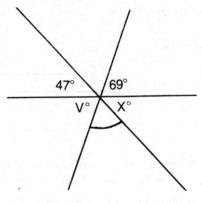

7.

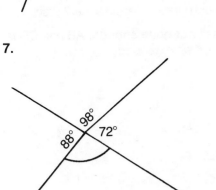

8.

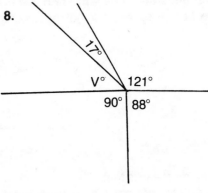

9.

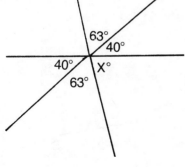

10.

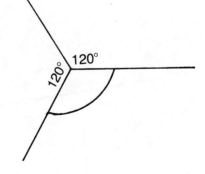

11.

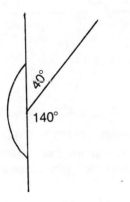

12.

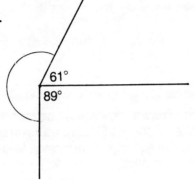

Parallel Lines

Exercise 15

The lines shown above are all **parallel** to one another. Measure each line, first in centimetres and then in inches. Record each result in your jotter. Should parallel lines be equal? Discuss.

Parallel lines are lines which do not meet. Now, this is **not quite enough**, AB and CD in **Figure 1** are lines which do not meet, but they would meet if we **extended** AB and CD.

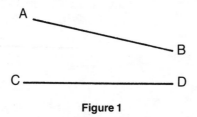

Figure 1

What we must say is that:
parallel lines are lines which never meet if we extend them in either direction.

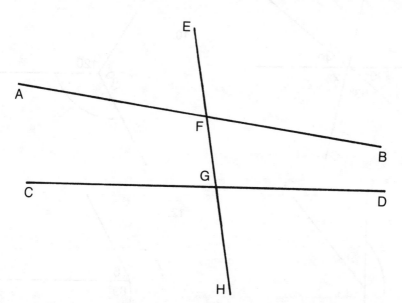

Figure 2

Look at **Figure 2**. Now you will see at once that AB is **not** parallel to CD. You do see that, don't you? We call the line EFGH a **transversal**. Find out the meaning of the word transversal. Copy the diagram above into your jotter. Measure all the angles in the figure. Make a note of each result in your jotter.

88

Now we'll give names to angles which lie in certain positions, when we have two lines (not necessarily parallel to each other), which are cut by a **transversal**.
Do you see that EFGH cuts **both** AB and CD in **Figure A**?
Look at the four angles with the x or y between their arms.
∠ BFG and ∠ DGH are **corresponding** angles. ∠ EFA and ∠ FGC are **corresponding angles**.

Name two other pairs of corresponding angles in the figure. So, when any two straight lines are cut by a transversal, we form four pairs of corresponding angles.
Now, if you know the sizes of two (**one** at **each** intersection point) of the eight angles in such a figure, you should be able to find the sizes of the remaining six angles.

Exercise 16

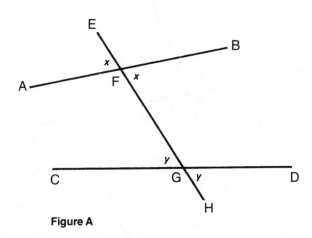

Figure A

1. In **Figure A** find the sizes of (a) ∠ EFB; (b) ∠ BFG; (c) ∠ AFG; if ∠ EFA = 74°.

2. Also find ∠ FGC; ∠ CGH and ∠ FGD; if ∠ DGH is 59°. **Do not use your protractor.**

3. Say which pairs of angles are **corresponding** and which pairs are **vertically opposite**.
   You will have found that you were able to work out the six angle sizes without using your protractor. All you need to know is summed up in the two diagrams, **Figure 1** and **Figure 2** below.

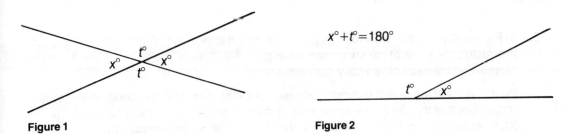

$x° + t° = 180°$

Figure 1

Figure 2

4. Now, in **Figure B** you have been given two lines (not parallel) cut by a transversal. Find the sizes of the six unmarked angles. Again, do not use your protractor. Name the pairs of **corresponding** angles and the pairs of **vertically opposite** angles.

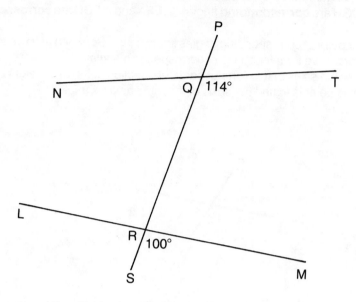

Figure B

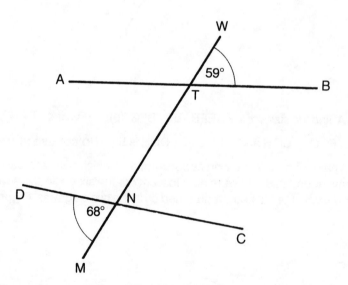

Figure C

5. In **Figure C** you have two lines, not parallel, cut by a transversal. Find, without using your protractor, the sizes of the six unmarked angles. Again name the pairs of corresponding angles and the pairs of vertically opposite angles.

Hint: A good method of reminding yourself of the position of corresponding angles is to remember that they lie on the **same side** of the transversal. When two lines are cut by the transversal, one angle lies **'inside'** and the other **'outside'** (see **Figure X**).

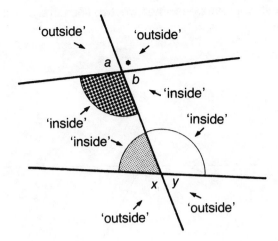

Figure X

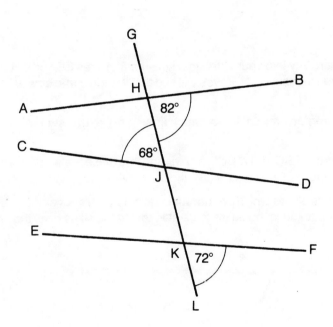

Figure Y

6. Of course, there need not be only two lines cut by the transversal. In **Figure Y** there are three lines, AB, CD and EF cut by the transversal GHJKL. Again, you have to find the sizes of the unmarked angles (nine in all). How would you describe the positions of corresponding angles made by these three lines and the transversal?

Now, we'll think about what happens when the two lines that are cut by the transversal are parallel, as in **Figure 1**.

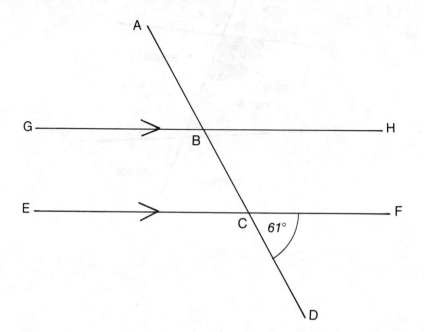

Figure 1

Copy the figure carefully into your jotter, making sure that lines EF and GH are parallel in your diagram. When you measure ∠ HBC you will find that it is the **same size** as ∠ FCD, that is 61°. So:

corresponding angles, made by parallel lines and a transversal, are equal to one another.

If you then measure ∠ ABG and ∠ BCE you will find that these are equal to each other, as if the case with ∠ ABH and ∠ BCF and as is also the case with ∠ GBC and ∠ ECD.

So if you are given two lines which are **parallel** and cut by a **transversal**, you need to be given the size of only one angle in the figure to be able to find the sizes of all the other angles in the figure.

Exercise 17

1. In **Figure 1** find the sizes of all the unmarked angles in the figure (seven in all). State which pairs are **corresponding** angles and which pairs are **vertically opposite** angles.

2. In **Figure 2** on page 93 AB and CD are parallel lines. EFGH is a transversal. ∠HGD = 109°. Find the sizes of all the unmarked angles in the figure (seven in all). State which pairs are corresponding angles and which pairs are vertically opposite angles.

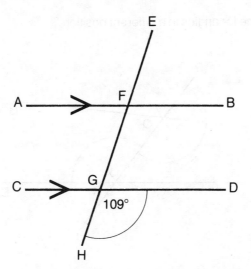

Figure 2

3. Now look at **Figure 3**. This figure looks a lot more difficult, doesn't it? We have **three** lines, which are parallel to one another, and two transversals cutting them. You have been given the sizes of two angles in the figure. Find the sizes of all the angles which are unmarked (twenty-two in all).

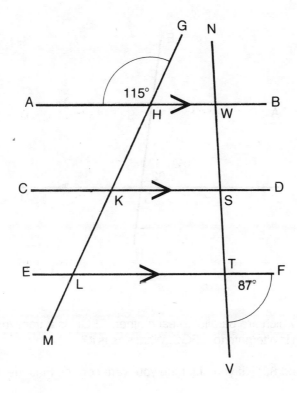

Figure 3

93

Now we'll talk about a name for angles in a different position.

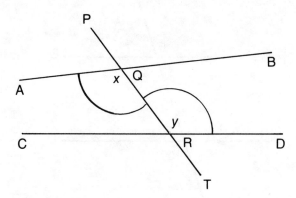

Again in the diagram above we have two lines (not parallel). The angles marked x and y lie on opposite sides of the transversal. Both lie 'inside' the figure; agreed? We call angles like these **alternate**. There is one other pair of alternate angles in the picture. Can you name them?

If the lines are parallel, the alternate angles made by the transversal are equal to each other.

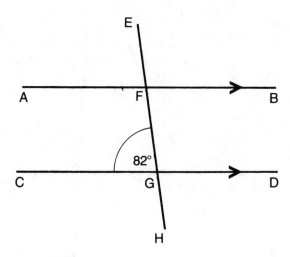

AB and CD are lines which are parallel to each other. EFGH is a transversal. ∠ FGC is 82°. Name the angle which is alternate to ∠ FGC. What size is it?

Did you say ∠ BFG and 82°? If you did, then you were correct. Find the size of all the other angles in the diagram.

Question

Name the six pairs of alternate angles in **Figure A**. If ∠ VDC = 81°, find the sizes of all the unmarked angles in the figure.

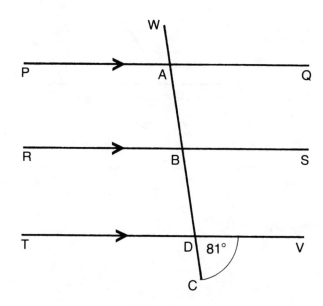

Figure A

Now we'll discuss what angles are called when they lie in the position we now show.

Look at the angles marked x and y in **Figure B**. These lie on the **same side** of the transversal. Both lie 'inside' the figure. We call these angles **co-interior**. If the lines are **not parallel** we cannot say anything about co-interior angles, except that they are co-interior.
However:

if the lines are parallel, the co-interior angles add up to 180°.

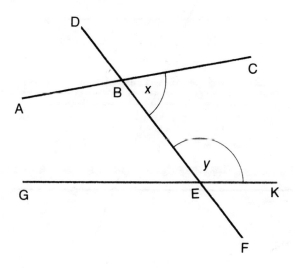

Figure B

Exercise 18

1. In **Figure 1**, AC and DF are parallel. ∠ CBE = 121°. Which angle is **co-interior** to ∠ CBE and what size is it? Name each angle in the figure and write its size beside it.

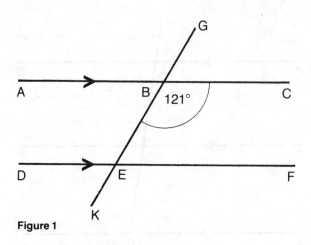

Figure 1

2. In **Figure 2**, AB, CD and EF are parallel lines. GKLMN is a transversal and ∠ KLD is 114°. Find the size of every angle in the figure.

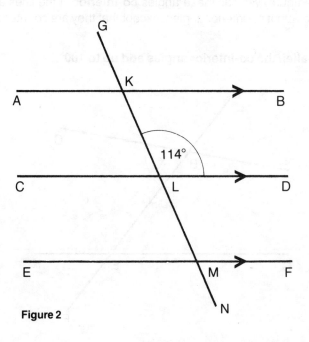

Figure 2

Exercise 19

Copy the following figures into your jotter. Find the sizes (without measurement) of the unmarked angles. **Name** the angles you find and list them in your jotter.

1.

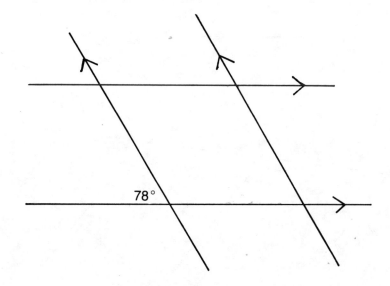

2.

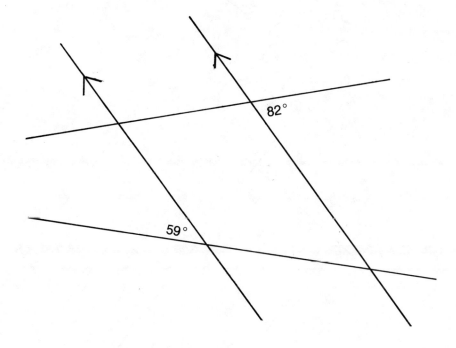

3.

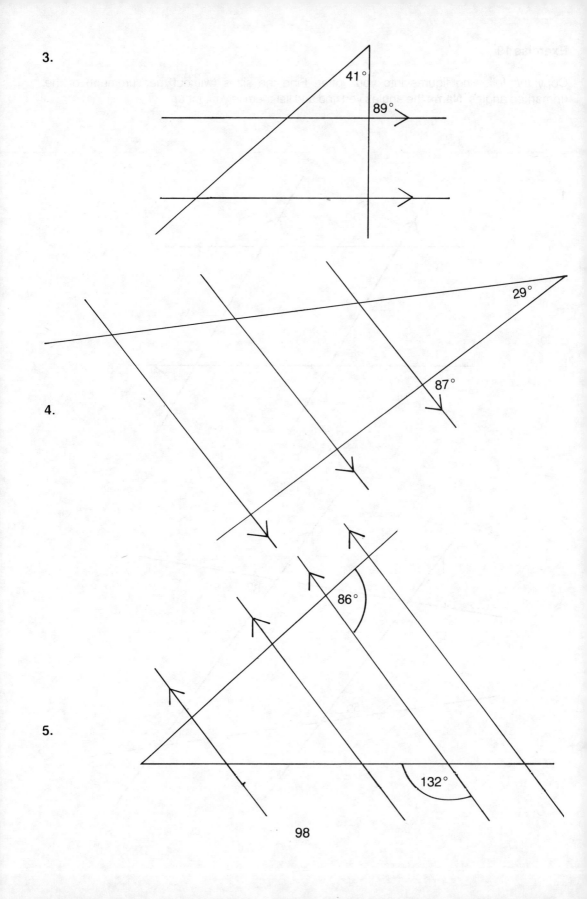

4.

5.

Triangles

We cannot draw a plane figure which is bounded by two straight lines. Do you agree with this?

We can draw **three** straight lines to enclose a plane figure.

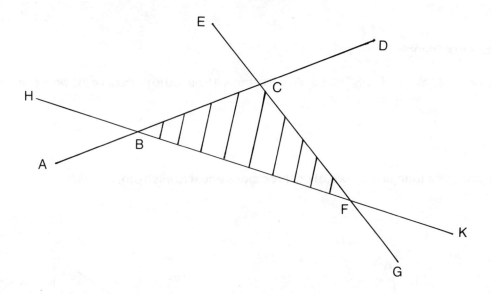

In the figure above, we have 3 straight lines enclosing a plane figure. We say that CBF is **a triangle**.

A triangle is a plane figure bounded by **three** straight lines.
There are several different kinds of triangles.

The Scalene Triangle

This is a triangle with its three sides all of **different lengths.** Its angles are all **different sizes.**
The triangle below is a **scalene triangle**.

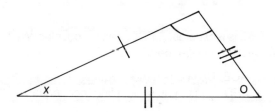

There is one thing about a scalene triangle that you can depend upon. Its angles, when added, will total 180°.

Exercise 20

In this exercise draw any triangle you like in your jotter. Take your protractor and carefully measure each angle. Write down your answers and check that the 3 angles total 180°. Remember, you will probably not get **exactly** 180°.

Isosceles Triangle

Look at the spelling of **ISOSCELES**. Your teacher will help you to pronounce the word correctly.

Exercise 21

An isosceles triangle has two of its three sides equal to each other.

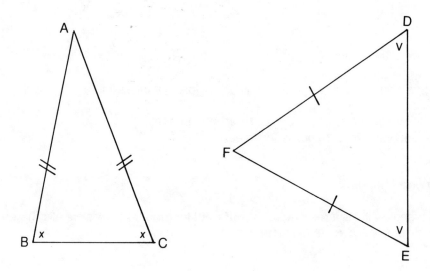

In the diagrams above, △ABC is isosceles. △FDE is also isosceles. We have marked the equal sides of each triangle with the 'same little mark'.

Now measure ∠ABC and ∠ACB. Do you find them equal?
Now measure ∠FDE and ∠FED. Again, you should find them equal.

Do you notice that the sets of equal angles are opposite the equal sides of the triangle?

In each of the following diagrams name the two equal angles and find their sizes. You should not need to **measure** the angles. Remember, the three angles of a triangle total 180°.

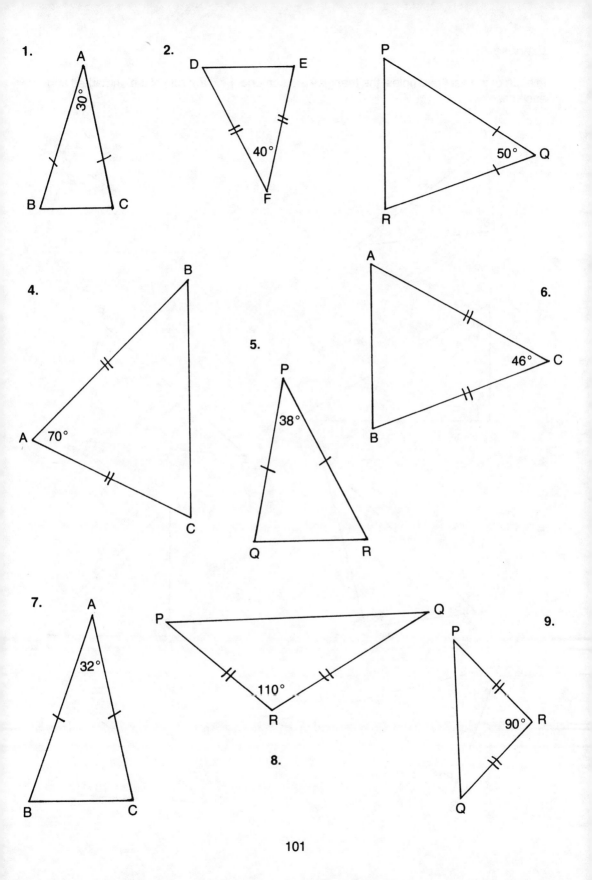

1.

2.

4.

5.

6.

7.

8.

9.

101

Exercise 22

In each of the next examples, the triangles are **scalene.** Find the size of the unmarked angle in each case.

1.

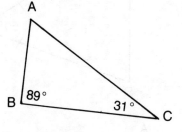

2.

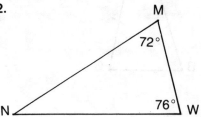

3.

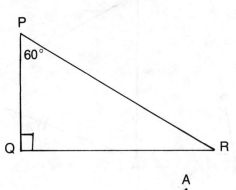

4.

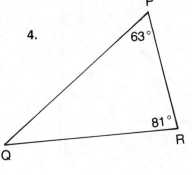

5.

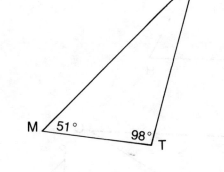

6.

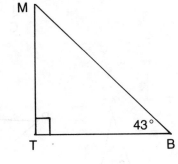

7.

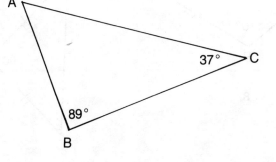

8.

9.

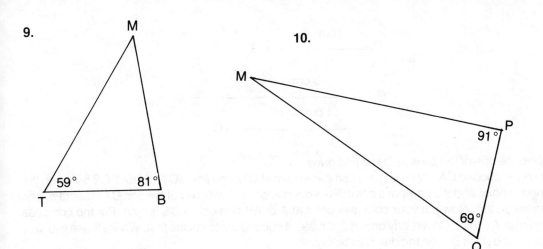

10.

Equilateral Triangle

When a triangle has all three sides equal in length the triangle is said to be equilateral.
When the three angles of a triangle are equal to one another we call the triangle equiangular.
Each angle is therefore 60°. Do you see why?

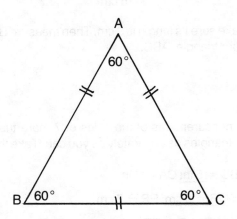

△ABC in the diagram above is an equilateral triangle and also an equiangular triangle.
Indeed, if it is one, it must be the other.

Construction of Triangles

Suppose we are given three straight lines and told to draw a triangle with these lengths as sides.
Suppose the three lines are AB = 3 cm, CA = 5 cm and BC = 4 cm. You want to draw △ABC.

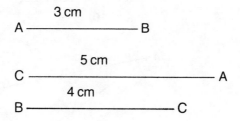

Choose one of the lines as **base** and draw it.

We will choose CA. Now put your compass point at C on the line BC and 'hold' CB between the pencil point and the compass point. Put your compass point on C of the line CA you draw, and draw an arc. Now put your compass point at A of AB and hold AB's length. Put the compass point at A on line CA already drawn and make a second arc to cut the first. Where those two arcs cut is the point B. See the diagram below.

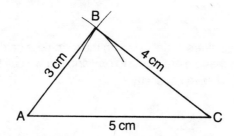

Now measure BA and make sure its length is 3 cm. Then measure BC and make sure its length is 4 cm. You now have your triangle ABC.

Exercise 23

Here you are given the measurements of the sides of 7 more triangles. Use your ruler and compasses and draw the triangles as accurately as you can. Take the underlined side as base.

1. $\triangle$ABC: AB = 3 in; $\underline{BC = 5\ in}$; CA = 4 in.

2. $\triangle$PQR: PQ = 5 cm; $\underline{QR = 6\ cm}$; RP = 7 cm.

3. $\triangle$DEF: DE = 2·5 in; $\underline{EF = 2\ in}$; FD = 2·5 in.
    What can you say about $\triangle$DEF?

4. $\triangle$PEK: PE = 2·4 in; $\underline{EK = 3·5\ in}$; KP = 2·4 in.
    What kind of triangle is $\triangle$PEK?

5. $\triangle$LMN: LM = 5 cm; $\underline{MN = 5\ cm}$; NL = 5 cm.
    What kind of triangle is $\triangle$LMN?

6. $\triangle$GHK: GH = 2·2 in; $\underline{HK = 2·2\ in}$; KG = 2·2 in.
    What kind of triangle is $\triangle$GHK?

7. $\triangle$RSV: RS = 1 in; $\underline{SV = 3\ in}$; VR = 2· 5 in.

Exercise 24

Now, using all the information that you have learned so far, try the following questions.
When there is only one angle at a vertex the angle is often named by the letter at that vertex.

1. In △ABC (**Figure 1**) AB = AC and ∠A = 100°. Find ∠C.

2. QR, a side of △PQR (**Figure 2**) has been **extended** to S. Find ∠PRS. Which is the greatest side of △PQR.
   Hint: The greatest side is opposite the greatest angle.

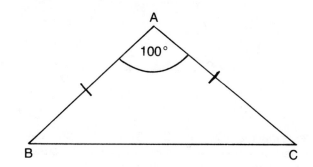

Figure 1

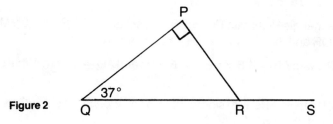

Figure 2

3. In **Figure 3**, find ∠D.
   After you have done this find ∠DFG.

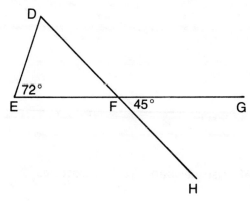

Figure 3

105

4. In **Figure 4**, find ∠X and ∠Z.

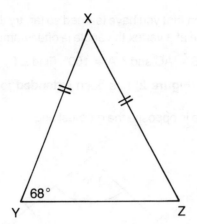

Figure 4

5. Draw an isosceles triangle ABC with AB = AC = 6 cm and ∠BAC = 74°. Make **a rough sketch first** to help you to see sides and angles in their correct positions. Measure the length of the side BC and the size of ∠ABC and ∠ACB.

6. Draw an equilateral triangle of side length 6·5 cm. Name your triangle PQR and say what you can about ∠P, ∠Q and ∠R.

7. Draw an isosceles triangle PQR with PQ = PR = 8 cm and ∠QPR = 100°. Measure the length of the side QR and the size of ∠PQR and ∠PRQ.

8. Draw a triangle ABC with AB = 4·8 cm; BC = 6·3 cm and AC = 6·2 cm. What can you say about ∠ABC?

9. Draw an equilateral triangle ABC with side length 5·5 cm. Extend the side BC to D and state, without measurement, the size of ∠ACD.

10.

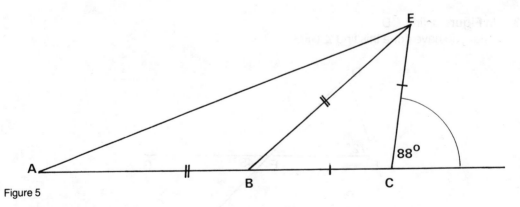

Figure 5

In figure 5 above, △BAE is isosceles and △ECB is isosceles.
Find ∠BEC and ∠AEC.

106

Right-angled Triangle

A right-angled triangle has one of its angles equal to 90°.

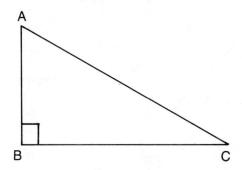

In the diagram above △ABC is a right-angled triangle. Why must ∠B, which is the right-angle, be the largest angle in the triangle?
Since ∠B is the largest angle, AC will be the longest side.
We call the longest side of a right-angled triangle **the hypotenuse**.

Exercise 25
Find the remaining angle in each of the triangles.

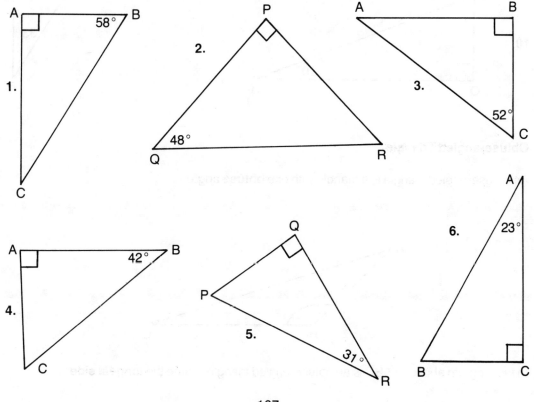

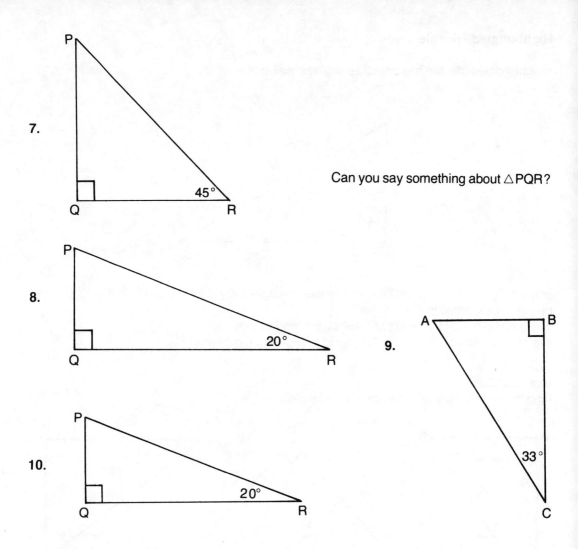

7.

Can you say something about △PQR?

8.

9.

10.

Obtuse-angled Triangle

An obtuse-angled triangle is a triangle with one **obtuse angle**.

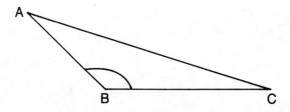

In the diagram above, △ABC is an obtuse-angled triangle. Name the **longest side**.

Acute-angled Triangle

An acute-angled triangle is a triangle with all three of its angles less than 90°. Think about this.
Must an acute-angled triangle be a scalene triangle?
Illustrate your answer by means of a figure.
We know now how to draw a triangle when we have been given the three side lengths. Now we are going to practise drawing a triangle when we know the lengths of two sides and the size of the angle made by these two sides.
Look at the diagram below.

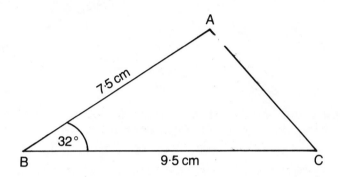

∠B is the angle made by the sides AB and BC whose lengths you have been given. We call such an angle an **included** angle. Included means 'caught in by'. In this case the angle has been 'caught in by' the two sides whose lengths you have been given.

Draw BC of length 9·5 cm. Put your protractor into position ready to measure, at B, an angle of 32°. Draw the arm, BA, of this angle of 32° and make the length of the arm BA equal to 7·5 cm. Join AC. You now have your triangle ABC.

Exercise 26

From the following data construct the triangle, naming it with the given letters.

1. ∠Q = 40°; QR = 4 cm; QP = 3 cm.
2. ∠N = 50°; NT = 10 cm; NM = 6 cm.
3. ∠Y = 32°; YZ = 8 cm; YX = 6·5 cm.
4. ∠P = 41°; PQ = 7 cm; PT = 4·5 cm.
5. ∠A = 50°; AR = 6 cm; AM = 4 cm.

6. ∠B = 25°; AB = 3·8 cm; BC = 2·9 cm.
7. ∠A = 49°; AB = 10 cm; AC = 8·5 cm.
8. ∠C = 106°; AC = 4·6 cm; BC = 8·2 cm.
9. ∠N = 90°; MN = 4·8 cm; NT = 6·2 cm.
10. ∠M = 104°; MN = 5·2 cm; MT = 4·8 cm.

Let's now think about drawing a triangle where we have been given the sizes of two angles and the length of one side. Such a triangle as PQR below for example.

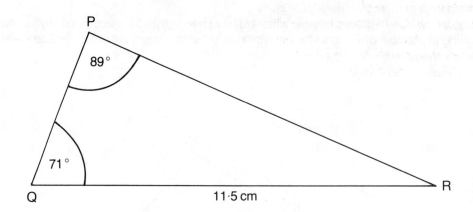

Now, can you see if we draw the line QR first (and after all, QR is the only line whose length we know), we will be able to draw the arm QP since we know that ∠Q = 71°, but we don't know where QP should stop.

To avoid this difficulty we work out the size of the angle at the other end of the line whose length we know (QR). We can do this because we know that the three angles of our triangle add up to 180°.

∠R = 180° − 71° − 89° = 20°.

Draw QR first, a length of 11·5 cm. At the end Q, make an angle of 71°. At the end R, make an angle of 20°. Position P is where the arms of these angles intersect. You now have your triangle, PQR.

Sometimes the data will provide you with a side length and an angle size at each end of that side length. If so, you can begin at once. If not, you must work out the size of the third angle of the triangle before you begin.

Exercise 27

In each of the following, draw the triangle, naming it from the given data.

1. BC = 8 cm; ∠B = 28°; ∠A = 92°.
2. PR = 6 cm; ∠P = 92°; ∠R = 35°.
3. AB = 5 cm; ∠A = 72°; ∠B = 64°.
4. AR = 10 cm; ∠A = 52°; ∠P = 72°.
5. RN = 4 cm; ∠R = 82°; ∠A = 41°.
6. MT = 6 cm; ∠M = 51°; ∠T = 42°.
7. BC = 7 cm; ∠C = 59°; ∠A = 74°.
8. AR = 3 cm; ∠R = 84°; ∠N = 28°.
9. HK = 8·5 cm; ∠H = 41°; ∠K = 58°.
10. TP = 7·5 cm; ∠T = 69°; ∠Q = 48°.

Properties of the Rectangle and Square

We all know what a rectangle looks like. We also all know what a square looks like. Of course, this is because so many everyday things are one or other of those shapes.

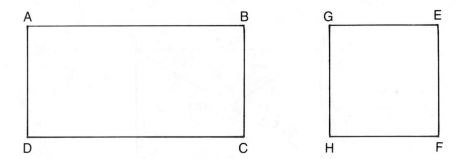

ABCD is a rectangle; GEFH is a square.
We have called this section 'properties' of the rectangle and square. Now property is something belonging to someone or to something. So the 'properties' of a rectangle are the things or facts that belong to the rectangle because it is a rectangle.

A rectangle has its opposite sides equal to one another.
In the diagram AB = DC and AD = BC. Those, then, are two properties of the rectangle.

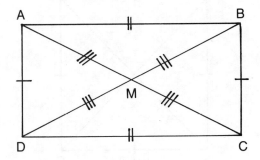

The four angles of a rectangle are right angles.
Again, that is one of the properties of the rectangle.
To make a diagonal we join two opposite vertices.
A rectangle has two diagonals and they are equal in length.
There you have another of the properties of the rectangle.
The diagonals of the rectangle bisect each other.
So the four line segments, AM, MC, DM and MB are of equal length. There you have another of the properties of the rectangle.
The angles round M are not right angles.

Now try to make a list of the properties of a square. In particular, what can you say about the angles round the point where the diagonals of a square intersect.
Draw a square and, make a list of its properties. Name your square PQRS, and let the four sides each be 6 centimetres long and let the diagonals intersect at T.

111

Exercise 28

1. ABCD is a rectangle. List the sides, angles and line segments which are not marked with a size and state the size. Give also (**a**) the area and (**b**) the perimeter of the rectangle.

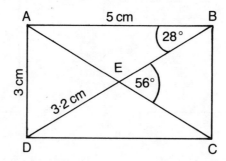

2. PQRS is a square (not drawn accurately). List the sizes of the unmarked lines and angles. Give also (**a**) the area and (**b**) the perimeter of the square.

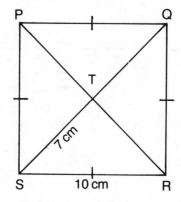

3. Now draw a rectangle, length 7 cm and breadth 5 cm. With as few measurements as possible, list the sizes and sides, line segments and angles in the figure when the diagonals are drawn.

   (**a**) Give the area and the perimeter of this rectangle.

   (**b**) Repeat the question for a rectangle of length 10 cm and breadth 8 cm.

   (**c**) Repeat the question for a square of side 11 cm.

   (**d**) Repeat the question for a square of side 8 cm.

4. Draw a rectangle, PQRS with PQ = 3 cm, and QR = 5·2 cm.
   Find its perimeter and its area.

5. Draw a square of side length 11·2 cm.
   Name your square ABCD. Find the perimeter and the area. Draw the lines AD and CB and measure each line.
   What can you say about AD and CB?

6. Draw a rectangle NMRT with NM = 6·9 cm and MR = 4·7 cm.
   Find the perimeter and the area.
   Draw NR and find the perimeter of △NMR.
   What can you say about the perimeter of △NRT **without measuring it**?

7. Draw a square ABCD with a **diagonal length** of 8 cm. Show clearly what steps you took and what measurements you made. What length is each side of your completed square?

8. Draw a rectangle whose area is 40 cm². You may **choose your own length and breadth**.
   Should all the members of the class have drawn exactly the same figure? Discuss this.

9. Draw a square with a perimeter of 20 cm. Name your square PQRT. Join PR and measure its length. Join TQ and let PR and TQ cross at A. Measure angle PAQ. What can you say about angles PAT, TAR and QAR **without measuring them**?

10. Draw a rectangle MNTQ of area 50 cm² and length 10 cm. State the breadth and calculate the perimeter length. Draw and measure each diagonal. Calculate the perimeter and the area of △MNT.

Up to this point in the book, we have only looked at, and thought about **plane** figures. We have studied many different plane figures, so far, and we said we would talk about plane figures at a later stage.

A plane figure is one which can be wholly drawn on a flat surface.

Any figure which you could draw on a page in your jotter would be a plane figure. Suppose you drew a shape, starting on the wall of the classroom, and you continued that shape, down and across the floor; the result would not be a plane figure, because it would not lie wholly on a flat surface. So, any plane figure is flat.

Now the shapes which we deal with in real life, take up a **volume** of space. Think of a house, a chair, a desk, ourselves. All of these occupy a certain volume. If we are good at art, we can draw figures on a flat surface in such a way that the figures 'look' solid. With plasticine or stone or clay, a model of the solid can be made, as we say, in **three dimensions.** It is quite difficult to show a solid on a flat surface. Think of the problems of showing areas of the earth on the surface of a map. Discuss with your teacher what makes this difficult to do.

Some of the solid shapes which we have learned to recognize have special names, such as cone, cylinder, sphere, cuboid, prism, etc. See how many of the shapes in Exercise 29 you can name. Can you explain why a drawing on a flat surface can be made to look three dimensional?

Exercise 29

Do you remember, for these figures, what the words 'face', 'edge' and 'vertex' mean? Discuss this. Copy and complete the table on page 115.

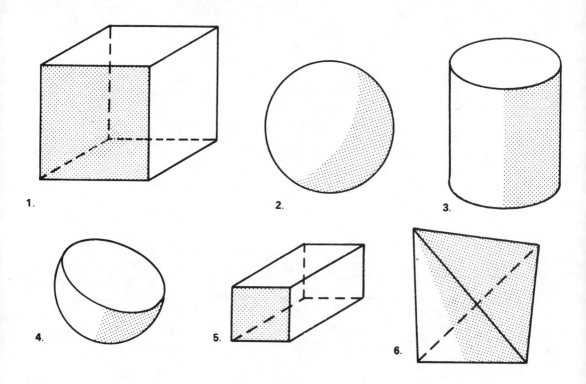

1.   2.   3.

4.   5.   6.

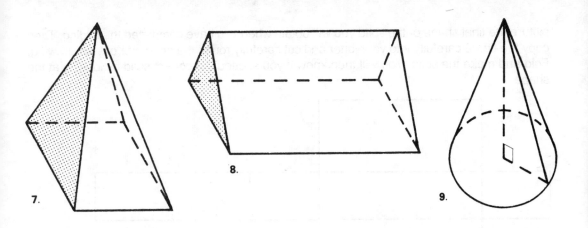

7.

8.

9.

	Number of faces	Number of edges	Number of vertices
Figure 1			
Figure 5			
Figure 6			
Figure 7			
Figure 8			

As a check on your result, try this test.

For each figure, add the number of faces to the number of vertices. Then subtract the number of edges. The result should be 2 in each case.

Exercise 30

Now try the following questions.

1. How many different lengths of edge has a cuboid?

2. How many different face sizes has a cuboid?

3. Is a cube always a cuboid?

4. Is a cuboid always a cube?

5. Try to make a list of objects in everyday use which are the shape drawn above.

Exercise 31

No doubt at some time you will have drawn certain shapes on cardboard, and by folding along certain lines, you will have been able to make cubes and cuboids and even more complicated figures. The flat shape you have before you fold is called the net of the three-dimensional figure you end up with. Now look at the nets drawn below. Each one is the net of some solid. Try to picture the final shape of the solid you will obtain when you have completed the folding. Then copy the figure carefully into your jotter and cut carefully round the perimeter of your drawing. Fold and make the solid. You will then know if you spotted the correct solid by looking at the shape.

1.

2.

3.

4.

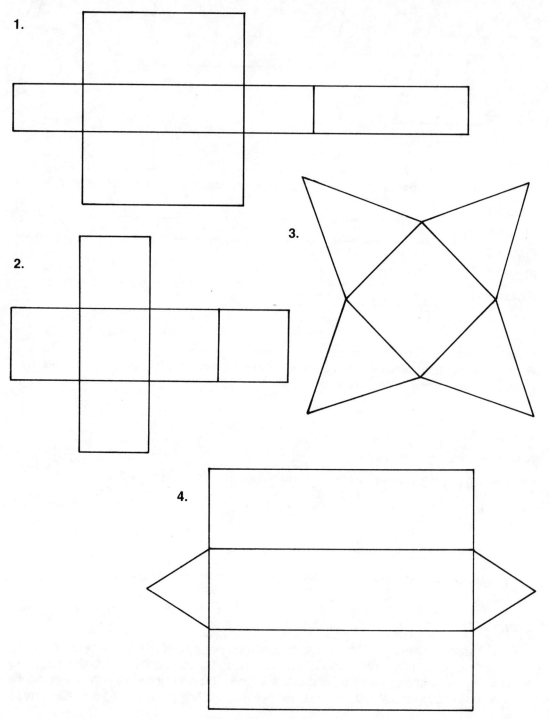

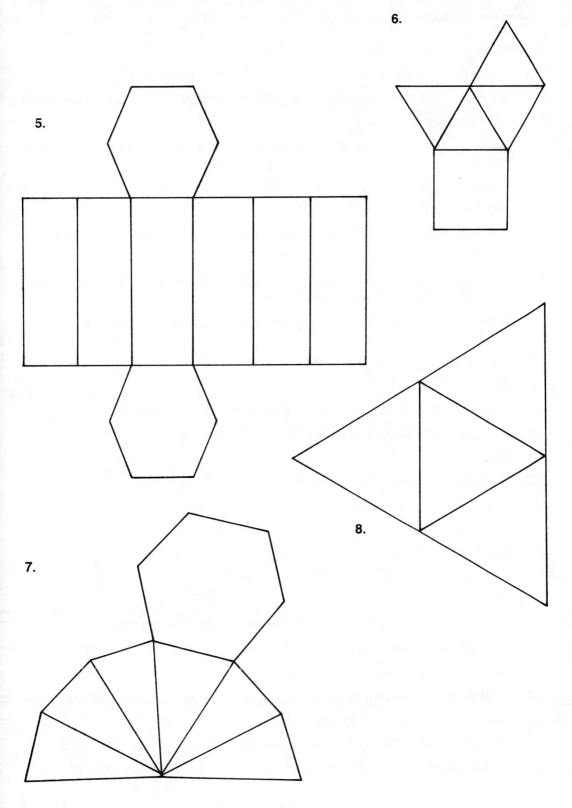

5.

6.

7.

8.

Miscellaneous Questions

Work answers to the following questions:

1. A man bought a car for £1,100 and sold it for £950. How much did he lose?

2. If 29 out of 10 000 people questioned do not watch sport on T.V., what percentage is this?

3. How much greater than $2\frac{1}{2}$ is $1\frac{3}{4} \times 2\frac{7}{8}$?

4. What interest has to be paid on £1,000 borrowed for 8 months at 12% per annum simple interest?

5. Which is smaller, $\frac{3}{7}$ or $\frac{5}{14}$?

6. What is the average of 15, 19, 23, 24 and 34.

7. An article which cost £4 had its price increased by 10%. What was the new price?

8. How many metres are there in 28·4 km?

9. An article costing £100 cash is sold on H.P. for a 20% deposit and 12 equal payments of £7·40. How much extra was charged for H.P.?

10. How many litres of water can a cuboid hold, if its internal measurements are 3 m, $2\frac{1}{2}$ m and $1\frac{3}{4}$ m?

11. If $a = 2$ and $x = 4$ what is the value of $2(3x + 5a)$?

12. Find x if $2x - 1 = 19$.

13. How many eggs can be bought for £1 if t eggs cost x pence altogether?

14. Give a simpler answer for $3a + 5a - 2a + 7a$.

15. Find t if $3t - 5 = 1$.

16. If q books cost £m how many pence does 1 book cost?

17. Give a simpler answer for $2p^2q \times 4q$.

18. Give a simpler answer for $\dfrac{5a^2b}{10a}$.

19. Find p if $p - \frac{1}{2} = 3$.

20. Write down the cost of c pencils if 1 pencil costs x pence. Give your answer in £ s

21. What is the supplement of 23°?

22. What is the complement of $84\frac{1}{2}$°?

23. What name is given to an angle of 90°?

24. How many degrees are in a straight angle?

25. State whether the following angles are acute, obtuse or reflex: 23°, 31°, 154°, 302°, 5°.

26. If two angles of a triangle are 21° and 102° respectively, what size is the third angle?

27. If an isosceles triangle has one of its equal angles equal to 40°, what are the sizes of the other two angles of the triangle?

28. What size is each angle of an equilateral triangle?

29. Draw two straight lines and a transversal and show a pair of alternate angles.

30. Draw two straight lines and a transversal and show a pair of corresponding angles.

31. Add 20% of £1 to 10% of 10 pence. Answer in pence.

32. Add $\frac{1}{2}$ of $2\frac{1}{2}$ to $5 \times \frac{3}{5}$.

33. Express $(2\cdot01 - 1\cdot86)$ as a percentage.

34. Write in figures, two million and four.

35. Write 1001 in words.

36. Multiply 2·34 by 10 and add your answer to $7\cdot8 \div 10$.

37. What interest would be due on a loan of £10,000 after 1 year at 15% simple interest?

38. Find the sum of 24, 326, 4028, 29 and 401.

39. Take 10% of £4 from 20% of £10. Answer in pence.

40. What fraction of 1 metre is 30 cm?

41. Write in simpler form, $12xy - xy + 3xy + x$.

42. Write in simpler form, $3x^2y \times 2x \times 4$.

43. Write in simpler form, $\dfrac{17a^2d}{34a}$.

44. Find d, if $3d = 81$.

45. Find t, if $\dfrac{t}{3} = 6$.

46. If a car travels x km in t hours, at what rate, in km/h, is it travelling?

47. Write down t% of £d.

48. Write in simpler form, $2t + 8t + 9t - 10t + 3$.

49. If $t = 3$, $p = 2$ and $q = 1$, find a number answer for $\dfrac{t^2 + p^2 + q}{21}$.

50. Find x if $23x + 1 = 27 - 3$.

51.

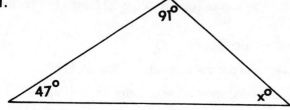

Find the value of x.

52.

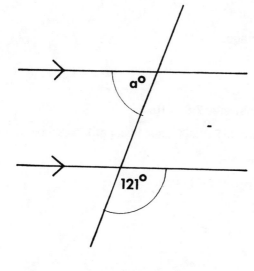

Find the value of a.

53.

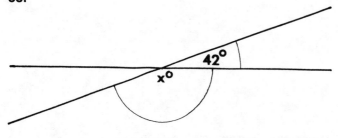

Find the value of x.

54.

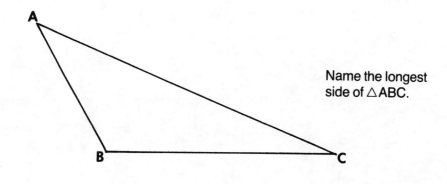

Name the longest side of △ABC.

55.

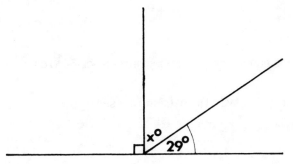

Find the value of *x*.

56.

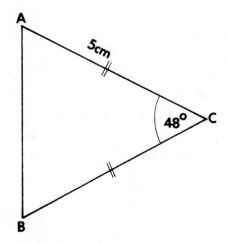

PQRS is a rectangle.
Find (a) the perimeter, in metres,
of rectangle PQRS
and (b) the area, in square metres,
of rectangle PQRS.

57.

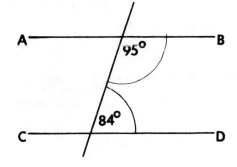

(a) State the length of BC.
(b) Find the size of ∠ABC.

58. What is the supplement of *p*°?

59. What is the complement of *t*°?

60.

What can you say about
the lines AB and CD?

61. Express as percentages: $\frac{2}{5}$, $\frac{1}{8}$ and $\frac{4}{5}$.

62. Find the average of 2, 3, 5, 9, 11 and 6.

63. What would have to be added to the sum of the six numbers in question 62 to give a total of 41?

64. A chair costing £129 is sold to gain 10% of what it cost. Find the selling price.

65. What should be taken from 11·29 to leave 7·345?

66. What must be added to $3\frac{1}{4}$ to make $7\frac{1}{8}$?

67. What percentage of £1 is 27 pence?

68. What fraction of £1 is 50 pence?

69. How many tiles, each a square of side length $\frac{1}{2}$ metre, will completely tile a floor which is rectangular with a length of 8 metres and a width of $6\frac{1}{2}$ metres?

70. Arrange $\frac{1}{2}$, $\frac{3}{5}$, $\frac{7}{10}$, $\frac{11}{20}$ and $\frac{3}{4}$ in ascending order.

71. If $t = 5$ and $x = 4$ find a number answer for $2(t^2 + x)$.

72. Find p, if $3p - 2 = 7$.

73. Find a simpler way of writing $\dfrac{12a + 24a - 3a}{11}$.

74. Write in simpler form $10a^2p \times 2ap^2$.

75. Write in simpler form $\dfrac{30x^2y}{10x}$.

76. If $a = 2$ and $c = 4$ write an equation connecting a and c.

77. What must be added to $2x$ to make $12x$?

78. What must be taken from $4xy$ to leave xy?

79. If $a = 7$, what is the value of $2a^2$?

80. Write in simpler form, $12p - p - p - p$.

81. What is the supplement of $2p°$?

82. What is the complement of $5x°$?

83.

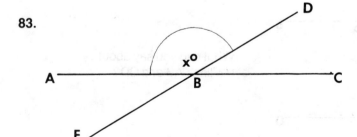

Name another angle in the figure which equals $x°$.

122

84. In the figure of question 83, what size is ∠DBC?

85.

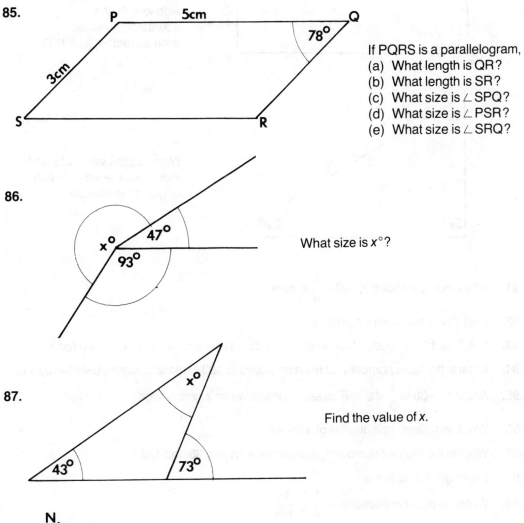

If PQRS is a parallelogram,
(a) What length is QR?
(b) What length is SR?
(c) What size is ∠SPQ?
(d) What size is ∠PSR?
(e) What size is ∠SRQ?

86.

What size is $x°$?

87.

Find the value of x.

88.

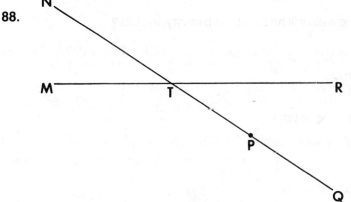

Name as many line segments as you can see in the figure above.

89.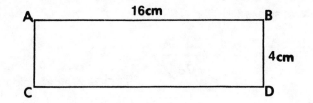

What would be the side length of a square of the same area as rectangle ABCD?

90.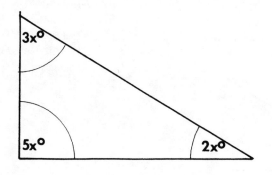

Work out the value of x and then state the size of each angle of the triangle.

91. What must be added to half of $\frac{4}{9}$ to make 1?

92. Find 2% of half a million pounds.

93. V.A.T. at 15% is added to a hotel bill of £250. What amount of V.A.T. is added?

94. What is the least number that must be added to 133 to make it exactly divisible by 11?

95. Add 2×3.08 to $\frac{1}{3}$ of 9.6. Express your answer as a percentage.

96. Write, in figures, one quarter of a million.

97. What is the biggest number that divides evenly into 33 and 132?

98. Express 24.2 kg in mg.

99. Write, as decimal fractions, $\frac{1}{2}, \frac{1}{4}, \frac{1}{8}, \frac{1}{10}$.

100. Tom was born in 1972. How old will he be, on his birthday, in 2001?

101. Find x if $\frac{3x}{10} = 30$.

102. Write in simpler form, $\frac{12mc^2}{6c}$.

103. Write in simpler form, $4p \times 2q \times 8p$.

104. Express t metres in cm.

105. Express x cm in km.

106. Write in simpler form, $13x - x + 11x$.

107. If $a = 1$ and $b = 2$ give a number to represent $2a + 3b$.

108. Find t if $2t - 27 = 3$.

109. Find p if $p - 14 = 14$.

110. Write in simpler form, $\dfrac{14p^2d}{28d^2}$.

111. What is the complement of $2\frac{1}{2}°$?

112. What kind of angle is an angle of $102°$?

113.

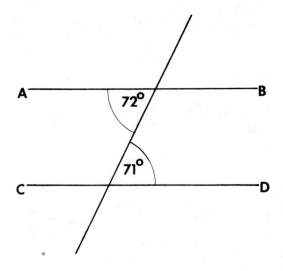

What can you say about the lines AB and CD?

114.

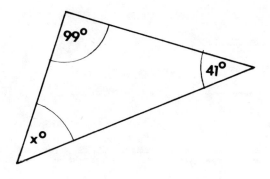

Find the value of x.

115.

Name a line which equals line BC + line CD.

116. What is the supplement of $142°$?

117.

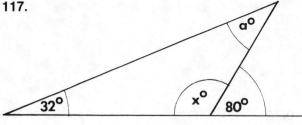

Find *a* and *x*.

118.

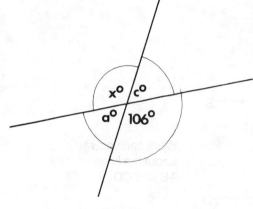

Find *x*, *c* and *a*.

119. Draw an isosceles triangle, ABC with AB = AC = 4·5 cm and ∠ BAC = 38°. Measure BC.

120.

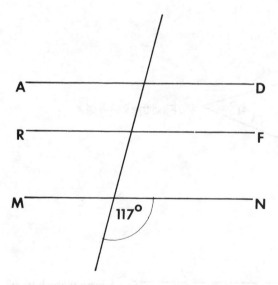

AD, RT and MN are parallel to one another.
Copy the figure and write in the sizes of all the un-marked angles.

121. Put into *descending* order, $\frac{1}{2}, \frac{1}{3}, \frac{3}{4}, \frac{7}{8}$ and $\frac{5}{16}$.

122. A 20% deposit is needed if an article costing £250 is to be bought on H.P. How much money is needed as deposit?

123. A boy's age is one-third that of his mother's. Their combined ages total 48 years. Find the age of each.

124. A jug of capacity $\frac{1}{2}$ litre is $\frac{2}{3}$ full. How many cm^3 could it still hold?

125. Four cubes each have a side length of 2 cm. How many more identical cubes would give a total volume of 56 cm^3?

126. Multiply 2 by $2\frac{1}{3}$ and subtract the result from 5.

127. Four boys are given £17·60 to divide equally among them. How much should each boy get?

128. What is the average of 1, 2, 3, 4, 5, 6 and 7?

129. Add 1·72 × 0·2 to 3 × 1·01.

130. Take the sum of $\frac{2}{3}$ and $\frac{4}{3}$ from 8.

131. Find p if $p + 18 = 35$.

132. If $a = 2$, what does $3a^2$ equal?

133. Write in simpler terms, $5t - 2t + 3t - t + 8$.

134. By how much is $12a$ greater than $3a$?

135. If $t = 5$ and $q = 4$ find a number answer for $2t + q^2$.

136. Multiply $2pq$ by $3q$.

137. Divide $28a^2$ by $7a$.

138. If $c = 5$, what does $24c$ equal?

139. Write in simpler form, $2ab + 12ab + 5ab + b$.

140. Find t, if $\frac{t}{8} = 5$.

141. How many degrees equal the sum of two right angles and a straight angle?

142. What is the supplement of $t°$?

143. What is the complement of 32°?

144. Do parallel lines ever meet?

145.

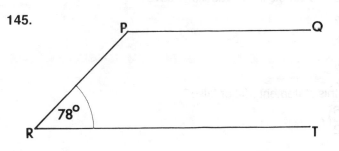

If PQ and RT are parallel lines what size is ∠QPR?

146. One angle of a triangle is 20°. A second angle is $t°$. What size is the third angle?

147. One of the diagonals of a rectangle measures 10 cm. What is the length of the other diagonal?

148. What can you say about the three sides of a scalene triangle?

149.

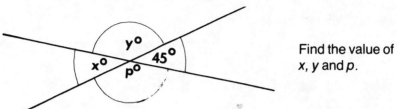

Find the value of x, y and p.

150.

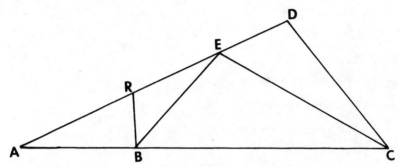

How many line segments can you name from the figure above?

151. Write, in figures, the number twenty-one thousand and six.

152. Write 2 000 004 in words.

153. A box of matches should contain 80 matches. If 10% of the matches have been used how many matches are left in the box?

154. What decimal fraction must be added to 0·079 to make 1?

155. Multiply 1·02 by 1000 and add 2·1 to your result.

156. What is the total surface area of the faces of a cuboid of length 3 metres of breadth 2 metres and of height 2 metres?

157. What is the volume, in cm³ of the cuboid in the question above?

158. What percentage of 40 is 8?

159. What fraction of 14 is 7?

160. What percentage of 14 is 7?

161. $a + b = 10$ if $a = b = 5$. Is this statement true or false?

162. If $t = 4$ what is the value of $\dfrac{5}{2}t$?

163. *a* and *b* are two numbers such that *a* is greater than *b* by 7. If *b* = 6, find *a*.

164. Find *a* if $2a - 17 = 9$.

165. Give a number answer for $\dfrac{2a}{c} + \dfrac{a}{d}$ if $a = 4, c = 8$ and $d = 1$.

166. Express $t\,m^3$ in cm^3.

167. Express $2cd$ pence in £'s.

168. Write in simpler form, $3p + 2p - p - 3p + 5p$.

169. The sum of $2a$ and $5a$ has $3a$ subtracted from it. What is the result?

170. If $a = 4$ and $t = 5$ what is the value of $2at^2$?

171. What is the supplement of $2p°$?

172.

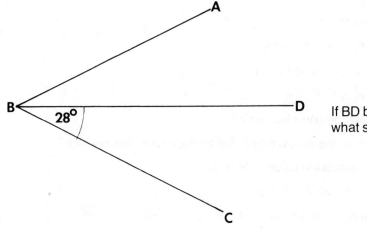

If BD bisects ∠ABC what size is ∠ABC?

173.

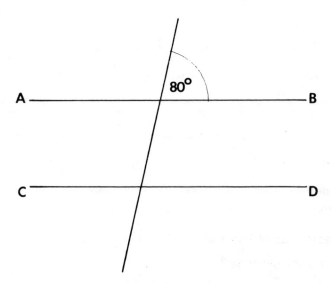

If AB and CD are parallel how many angles of 80° are there in the figure?

174. One angle of a triangle is 41°. A second angle is 32°. What size must the third angle be?

175. What name is given to an angle which is smaller than 90°?

176. How many degrees are in a complete revolution?

177. What name is given to an angle greater than 180° but smaller than 360°?

178. Draw $\triangle ABC$ with AB = 4·8 cm, BC = 6·2 cm and CA = 5·3 cm. Measure $\angle$ ABC.

179. Draw $\triangle PQR$ with PR = 4·2 cm, QR = 5·7 cm and $\angle$ PRQ = 39°. Measure PQ.

180. What is the complement of $m°$?

181. Add $2\frac{1}{2} \times \frac{1}{3}$ to $\frac{1}{6}$ and take the result from 2.

182. A jug holds $\frac{3}{4}$ litre. If it is half full of water how many cm^3 of water can it still hold?

183. Divide 2·04 by 0·02.

184. Multiply 4·2 by 0·01.

185. Add 1·2 × 10 to 0·03 × 100.

186. Take the sum of $1\frac{1}{2}$ and $3\frac{1}{2}$ from 10.

187. What percentage of 200 is 50?

188. What fraction of $4\frac{1}{2}$ is $1\frac{1}{2}$?

189. How many days are there in a leap year?

190. 1984 is a leap year and is divisible by 8. Are all leap years divisible by 8?

191. $c = 4$ and $t = 1$. What is the value of $2c + 3t$?

192. If $a = 5$ what is the value of a^3?

193. What value of x would make x^2 equal to x^3 ?

194. Find x, if $2x - 3 = 109$.

195. Find x, if $13x = 26$.

196. Find a, if $\dfrac{a}{2} + 1 = 5$.

197. Find y, if $\dfrac{y}{3} + 1 = 6$.

198. Find t, if $2t = 4\frac{1}{2}$.

199. Write, in simpler form, $2m \times 3m^2 \times 6$.

200. Write, in simpler form, $\dfrac{5pm^2}{10m}$.

201. What angle is three times the size of a right angle?

202. How many degrees are in a straight angle?

203.

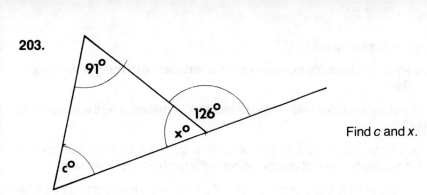

Find *c* and *x*.

204.

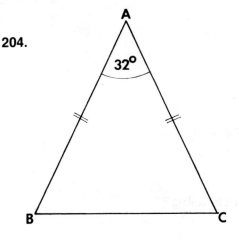

Find ∠ABC and ∠ACB.

205.

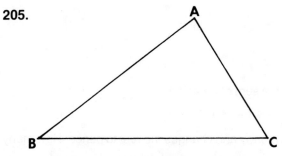

Why is the line BC smaller than the line BA added to the line AC?

206. In the question above if ∠B = 28° and ∠A is 111° what size is ∠C?

207.

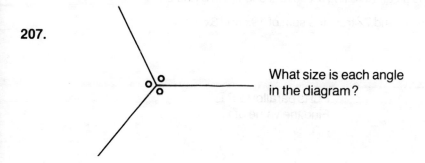

What size is each angle
in the diagram?

208. What name is given to an angle of 121°?

209. If one side of an equilateral triangle measures 6 cm what must be the sum of the lengths of the other two sides?

210. One of the angles of an isosceles triangle measures 110°. What must be the sizes of the two other angles?

211. It costs 80p to go to a Bingo session. Fifty ladies wish to go in a party to play Bingo. For every 10 ladies, 1 is admitted free of charge. What is the total charge for admission?

212. A boy has £2·45. He wishes to buy 3 pencils costing 8p each and a ruler costing 24p. If he does this how much change will he have?

213. What fraction of 4 is 1?

214. What percentage of 4 is $\frac{1}{2}$?

215. What fraction of 8 is $\frac{1}{4}$?

216. Express 0·08 as a percentage.

217. What must be added to the sum of $\frac{1}{3}$ and $\frac{1}{6}$ to give 1?

218. What must be taken from 2·824 to leave 0·06?

219. What percentage of 102 is 306?

220. A car travels 87 km in 3 hours. At what rate is it travelling?

221. Find p if $\dfrac{2p}{3} = 4$.

222. Find a if $2a - 17 = 33$.

223. Multiply $14ab$ by $2a^2$.

224. Divide $4m^2n$ by n^2.

225. Add $2p$ and $7p$. From their sum, take the sum of $2p$ and p.

226. Write, in simpler form, $2d + 8d - 6d + 4d - d + 5$.

227. A boy has £x. He gives away half his money and changes the rest to pence. How many pence has he?

228. Express $2c$ kg in mg.

229. At what speed is a car travelling if it covers $55d$ km in d hours?

230. Take the sum of $4x$ and $7x$ from the sum of $19x$ and $3x$.

231.

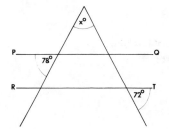

PQ is parallel to RT.
Find the value of x.

232.

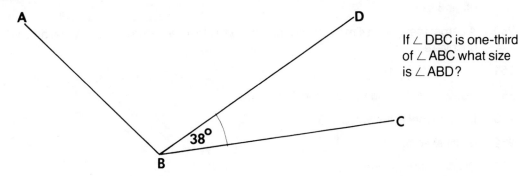

If ∠DBC is one-third of ∠ABC what size is ∠ABD?

233. What is the complement of $12t°$?

234. If $x°$ is the supplement of $y°$ write down an equation connecting x and y.

235. Draw a triangle whose angles are 32°, 49° and 99°. Is this triangle *unique*?

236. Draw △TMN with TM = 8 cm, MN = 6 cm and ∠TMN = 42°. Measure TN.

237. Draw a line 11·6 cm long. Call this line AB. Use your ruler and compasses to find the mid-point of AB *without actual measurement*.

238. If $2t°$ is the supplement of $p°$ write down an equation connecting t and p.

239. What name is given to an angle of 11°?

240.

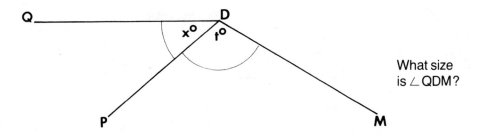

What size is ∠QDM?

241. Find the sum of 31·2, 3·04 and 26·12.

242. Express as fractions in lowest terms: 0·8, 0·6, 0·08 and 0·002.

243. Express 0·41 as the sum of two fractions.

244. Express as decimals: $\frac{7}{10}$, $\frac{6}{100}$, $\frac{3}{1000}$, $\frac{64}{100}$ and $\frac{235}{1000}$.

245. Divide $8\frac{2}{5}$ by $2\frac{5}{8}$.

246. Multiply $2\frac{2}{3}$ by $3\frac{1}{2}$ and divide your result by $2\frac{4}{5}$.

247. Express $3\frac{3}{8}$ as an improper fraction.

248. By a *short* method multiply 236 by 99.

249. The perimeter of a rectangle is 42·4 cm. One side is 5·1 cm long. Find the lengths of the other 3 sides.

250. A carpet measuring 4 m by 3 m is laid in a room with a floor area of 14 m². What area is left uncovered?

251. Find x, if $3x = 4\frac{1}{2}$.

252. Add $2a$ and $7a$ and take $3a$ from their sum.

253. If $t = 4$ what is the value of $5t^2$?

254. Multiply $4c$ by $2c^2$.

255. Divide $18mt^2$ by $9mt$.

256. Find c if $\frac{c}{2} + 1 = 3$.

257. Express t minutes in hours.

258. What is the number which is half of the number $14x$?

259. If $a = 2$ and $d = 3$ find a value for $2a + 7d$.

260. If $a = 5$ and $t = 4$ and $xat = 40$ find x.

261. One angle of a triangle is 38°. A second is 102°. Find the third.

262.

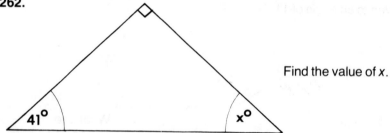

Find the value of x.

263.

Write down the length of CB.

264.

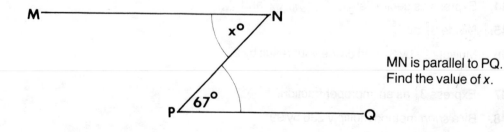

MN is parallel to PQ. Find the value of x.

265.

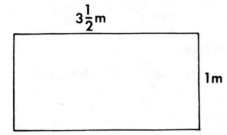

$3\frac{1}{2}$m

1m

The figure shown is a rectangle. Give its area in cm².

266. What is the complement of $\frac{1}{2}t°$.

267. What fraction of a straight angle is 20°?

268. What fraction of a right angle is $22\frac{1}{2}°$?

269.

A — D — B

If AB measures *x* cm and DB measures *y* cm write down the length of AD.

270. What name is given to an angle of 271°?

271. If 3 books each costing £1·75 are paid for from a £10 note what change will there be?

272. Multiply $2\frac{1}{4}$ by 8 and divide the result by $2\frac{1}{2}$.

273. Add 1·02, 2·03, 13·6 and 121.

274. Express $\frac{1}{4}$ of $\frac{1}{10}$ as a percentage.

275. Find the average of 8, 11, 17, 23 and 31.

276. What must be added to 1·2 × 2·4 to give 10·2?

277. Divide 2·04 by 0·02.

278. Multiply 23·1 by 0·01.

279. What interest is due on £100 borrowed for 6 months at 10% per annum simple interest?

280. Add 40% of £1000 to 20% of £3000.

281. Write, in simpler form, $24x - x + 2x + 9x + 7$.

282. Multiply 17*pq* by *ca*.

283. Divide 40*x* by 20.

284. Find *p* if $3p + 1 - 40$.

285. Express *m* days in hours.

286. Express *t* mg in kg.

287. Add 2*x*, 4*x* and 11*x* and take 12*x* from your result.

288. Find *m* if $\frac{m}{4} = 2\frac{1}{2}$.

289. Write, in simpler form, $\dfrac{12abc}{2bd} \times \dfrac{d}{ac}$.

290. What distance is travelled in p hours at x km/hour?

291. Draw $\triangle ABC$, with AB = 8·6 cm, BC = 9·8 cm and $\angle ABC = 29°$. Measure AC.

292. What is the complement of $3x°$?

293. What fraction of a complete revolution is 40°?

294.

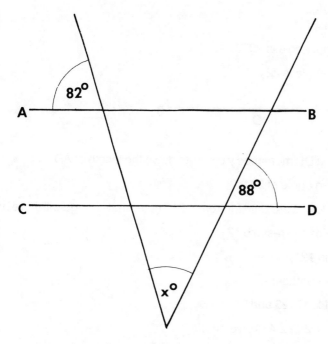

AB and CD are parallel lines. What is the value of x?

295. Draw an angle of 324°

296. Draw a rectangle whose diagonals are 10 cm long. Is this rectangle *unique*?

297. If the angles of a triangle are $a°$, $x°$ and $t°$, write an equation connecting a, x and t.

298.

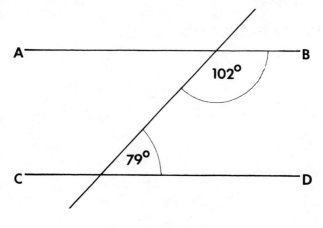

What can you say about the lines AB and CD?

299.

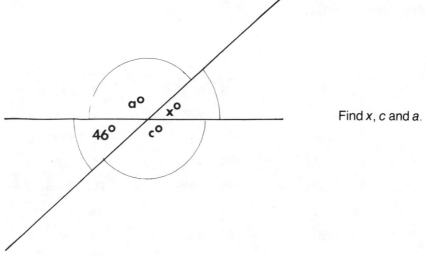

Find x, c and a.

300. Draw a triangle PQR with PQ = 8 cm, QR = 6·8 cm and PR = 8·2 cm. Measure ∠ PQR.

Answers to Arithmetic exercises

Exercise 1

1.	341	**2.**	180	**3.**	9211	**4.**	817
5.	297	**6.**	422	**7.**	930	**8.**	254
9.	192	**10.**	9695	**11.**	42	**12.**	1247
13.	1533	**14.**	7105	**15.**	1139	**16.**	200
17.	3058	**18.**	3835	**19.**	1089	**20.**	3043
21.	137	**22.**	908	**23.**	176	**24.**	26
25.	46	**26.**	1202	**27.**	680	**28.**	58
29.	454	**30.**	6562	**31.**	196	**32.**	137
33.	326	**34.**	183	**35.**	66	**36.**	111
37.	87	**38.**	197	**39.**	3000	**40.**	3935

Exercise 2

1.	343	**2.**	192	**3.**	230	**4.**	39
5.	2041	**6.**	1107	**7.**	377	**8.**	224
9.	3869	**10.**	1112	**11.**	1206	**12.**	918
13.	2206	**14.**	1872	**15.**	345	**16.**	123
17.	968	**18.**	140	**19.**	441	**20.**	6737
21.	6223	**22.**	888	**23.**	644	**24.**	306
25.	415	**26.**	46	**27.**	352	**28.**	222
29.	362	**30.**	238	**31.**	139	**32.**	422
33.	289	**34.**	255	**35.**	156	**36.**	138
37.	460	**38.**	156	**39.**	308	**40.**	79

Exercise 3

1.	726	**2.**	72	**3.**	28	**4.**	3162
5.	108	**6.**	415	**7.**	268	**8.**	19
9.	304	**10.**	763	**11.**	315	**12.**	442
13.	253	**14.**	892	**15.**	3093	**16.**	4845
17.	3291	**18.**	662	**19.**	1	**20.**	303

21.	16	22.	285	23.	17	24.	22
25.	201	26.	8	27.	14	28.	18
29.	1	30.	1	31.	3	32.	565
33.	127	34.	98	35.	203	36.	907
37.	2	38.	1	39.	897	40.	2

Exercise 4

1.	185	2.	532	3.	225	4.	85
5.	9587	6.	432	7.	450	8.	270
9.	23	10.	18	11.	58	12.	4997
13.	68	14.	274	15.	1857	16.	9048
17.	68	18.	32	19.	1	20.	3
21.	1	22.	62	23.	4	24.	99
25.	13	26.	3093	27.	45	28.	4
29.	33	30.	1	31.	3	32.	4
33.	725	34.	7	35.	3	36.	29
37.	12	38.	0	39.	198	40.	1053

Exercise 5

1.	92	2.	3056	3.	588	4.	3036
5.	1812	6.	2478	7.	1344	8.	2169
9.	2198	10.	984	11.	1040	12.	207
13.	3744	14.	13184	15.	891	16.	14200
17.	2664	18.	9648	19.	344	20.	2976
21.	0	22.	1208	23.	57	24.	510
25.	36	26.	752	27.	268	28.	117
29.	5136	30.	1905	31.	492	32.	4992
33.	68	34.	573	35.	54	36.	1111
37.	1740	38.	297	39.	2008	40.	6173

Exercise 6

| 1. | 144 | 2. | 432 | 3. | 96 | 4. | 84 |

5. 1080	**6.** 308	**7.** 528	**8.** 256
9. 588	**10.** 126	**11.** 840	**12.** 160
13. 264	**14.** 915	**15.** 704	**16.** 16
17. 84	**18.** 1202	**19.** 1848	**20.** 448
21. 144	**22.** 192	**23.** 216	**24.** 68
25. 192	**26.** 50	**27.** 15	**28.** 160
29. 96	**30.** 204	**31.** 240	**32.** 608
33. 20	**34.** 144	**35.** 72	**36.** 144
37. 144	**38.** 64	**39.** 1024	**40.** 384

Exercise 7

1. 943	**2.** 768	**3.** 3450	**4.** 308
5. 598	**6.** 986	**7.** 8217	**8.** 462
9. 576	**10.** 441	**11.** 1024	**12.** 899
13. 504	**14.** 1232	**15.** 416	**16.** 2378
17. 253	**18.** 700	**19.** 483	**20.** 648
21. 324	**22.** 342	**23.** 288	**24.** 256
25. 121	**26.** 152	**27.** 399	**28.** 1056
29. 1786	**30.** 1148	**31.** 378	**32.** 513
33. 1152	**34.** 1488	**35.** 899	**36.** 864
37. 558	**38.** 1189	**39.** 192	**40.** 196

Exercise 8

1. 5R3	**2.** 381	**3.** 56R1	**4.** 86R4
5. 196R8	**6.** 10R4	**7.** 26R2	**8.** 242R10
9. 422R2	**10.** 8R1	**11.** 141	**12.** 2R4
13. 48R7	**14.** 5R4	**15.** 38R2	**16.** 53R2
17. 13R4	**18.** 138R3	**19.** 96	**20.** 103R3
21. 5R2	**22.** 3R2	**23.** 80R1	**24.** 3R3
25. 2	**26.** 5R1	**27.** 4R3	**28.** 60R1
29. 9R6	**30.** 2R6	**31.** 4R7	**32.** 15R1
33. 9R2	**34.** 7R5	**35.** 29	**36.** 9R3
37. 2	**38.** 13R1	**39.** 12	**40.** 20R4

Exercise 9

1. 4R2	**2.** 15R4	**3.** 8	**4.** 20R3
5. 11R7	**6.** 2R6	**7.** 11R4	**8.** 9R3
9. 4R1	**10.** 8R1	**11.** 43R2	**12.** 8R6
13. 46R1	**14.** 3R3	**15.** 9R3	**16.** 12R2
17. 0R1	**18.** 1R1	**19.** 1R2	**20.** 18R6

Exercise 10

1. 6R1	**2.** 29R2	**3.** 21R1	**4.** 7R2
5. 11R1	**6.** 6R1	**7.** 2R5	**8.** 8R4
9. 14R3	**10.** 16R7	**11.** 30R1	**12.** 50
13. 67R5	**14.** 7	**15.** 35R5	**16.** 42
17. 52R5	**18.** 111R1	**19.** 6R6	**20.** 45R1

Exercise 11

1. 16	**2.** 26	**3.** 11	**4.** 37
5. 11	**6.** 36	**7.** 32	**8.** 104
9. 17	**10.** 26	**11.** 64	**12.** 50
13. 17	**14.** 49	**15.** 56	**16.** 21
17. 56	**18.** 8	**19.** 70	**20.** 156

Exercise 12

1. 23	**2.** 21	**3.** 37	**4.** 14
5. 22	**6.** 82	**7.** 21	**8.** 314
9. 98	**10.** 110	**11.** 712	**12.** 54
13. 183	**14.** 4	**15.** 66	**16.** 91
17. 730	**18.** 21	**19.** 1343	**20.** 178

Exercise 13

1. 13	**2.** 13	**3.** 19	**4.** 34
5. 25	**6.** 45	**7.** 10	**8.** 22
9. 16	**10.** 67	**11.** 135	**12.** 14

13. 3	**14.** 8	**15.** 22	**16.** 37
17. 21	**18.** 20	**19.** 13	**20.** $33\frac{1}{2}$

Exercise 14

1. 52	**2.** 20	**3.** 6	**4.** 21
5. 27	**6.** 7	**7.** 35	**8.** 2
9. 7	**10.** 10	**11.** 2	**12.** 0
13. 0	**14.** 0	**15.** 583	**16.** 20
17. 30	**18.** 73	**19.** 20	**20.** 1

Exercise 16

1. 2	**2.** 4	**3.** 6	**4.** 8
5. 10	**6.** 3	**7.** 5	**8.** 7
9. 9	**10.** 11	**11.** 13	**12.** 15
13. 17	**14.** 19	**15.** 21	**16.** 12
17. 14	**18.** 16	**19.** 18.	**20.** 20

Exercise 17

1. 4	**2.** 8	**3.** 12	**4.** 16
5. 20	**6.** 5	**7.** 9	**8.** 13
9. 17	**10.** 21	**11.** 6	**12.** 10
13. 14	**14.** 18	**15.** 22	**16.** 7
17. 11	**18.** 15	**19.** 19	**20.** 23

Exercise 18

1. 8	**2.** 16	**3.** 24	**4.** 32
5. 40	**6.** 12	**7.** 20	**8.** 28
9. 36	**10.** 44	**11.** 10	**12.** 18
13. 26	**14.** 34	**15.** 42	**16.** 14
17. 22	**18.** 30	**19.** 38	**20.** 46

Exercise 19

1. 1	**2.** 4	**3.** 16	**4.** 3
5. 12	**6.** 100	**7.** 4	**8.** 10
9. 2	**10.** 36	**11.** 21	**12.** 32
13. 12	**14.** 15	**15.** 10	**16.** 15
17. 1	**18.** 15	**19.** 72	**20.** 14

Exercise 20

1. $\frac{3}{4}$	**2.** $\frac{4}{5}$	**3.** $\frac{1}{3}$	**4.** $\frac{1}{8}$
5. $\frac{1}{3}$	**6.** $\frac{1}{4}$	**7.** $\frac{1}{2}$	**8.** $\frac{3}{4}$
9. $\frac{1}{2}$	**10.** $\frac{1}{3}$	**11.** $\frac{1}{2}$	**12.** $\frac{2}{3}$
13. $\frac{1}{4}$	**14.** $\frac{1}{2}$	**15.** $\frac{1}{2}$	**16.** $\frac{1}{6}$
17. $\frac{1}{4}$	**18.** $\frac{1}{3}$	**19.** $\frac{1}{2}$	**20.** $\frac{2}{3}$
21. $\frac{1}{2}$	**22.** $\frac{3}{4}$	**23.** $\frac{7}{8}$	**24.** $\frac{1}{4}$
25. $\frac{3}{8}$	**26.** $\frac{1}{4}$	**27.** $\frac{1}{2}$	**28.** $\frac{2}{5}$
29. $\frac{3}{5}$	**30.** $\frac{1}{50}$	**31.** $\frac{1}{16}$	**32.** $\frac{1}{12}$
33. $\frac{1}{8}$	**34.** $\frac{1}{6}$	**35.** $\frac{5}{24}$	**36.** $\frac{1}{4}$
37. $\frac{1}{3}$	**38.** $\frac{1}{5}$	**39.** $\frac{1}{3}$	**40.** $\frac{1}{4}$

Exercise 21

1. $2\frac{1}{4}$	**2.** $1\frac{3}{5}$	**3.** $1\frac{2}{3}$	**4.** $2\frac{2}{3}$
5. $2\frac{1}{3}$	**6.** $3\frac{1}{3}$	**7.** $3\frac{2}{3}$	**8.** $4\frac{2}{3}$
9. $1\frac{1}{4}$	**10.** $1\frac{3}{4}$	**11.** $2\frac{1}{4}$	**12.** $3\frac{1}{4}$
13. $4\frac{1}{4}$	**14.** $1\frac{4}{5}$	**15.** $2\frac{1}{5}$	**16.** $2\frac{2}{5}$
17. $3\frac{3}{5}$	**18.** $4\frac{1}{5}$	**19.** $2\frac{1}{6}$	**20.** $1\frac{1}{6}$
21. $3\frac{1}{6}$	**22.** $2\frac{5}{6}$	**23.** $3\frac{2}{3}$	**24.** $1\frac{1}{8}$
25. $1\frac{3}{8}$	**26.** $1\frac{5}{8}$	**27.** $2\frac{3}{8}$	**28.** $2\frac{5}{8}$
29. $2\frac{1}{10}$	**30.** $3\frac{1}{10}$	**31.** $4\frac{3}{10}$	**32.** $2\frac{7}{8}$

33. $3\frac{7}{8}$ **34.** $5\frac{1}{10}$ **35.** $6\frac{3}{10}$ **36.** $7\frac{3}{10}$

37. $3\frac{3}{8}$ **38.** $3\frac{9}{10}$ **39.** $5\frac{4}{5}$ **40.** $5\frac{3}{4}$

Exercise 22

1. $\frac{15}{4}$ **2.** $\frac{9}{4}$ **3.** $\frac{7}{4}$ **4.** $\frac{19}{4}$

5. $\frac{21}{4}$ **6.** $\frac{27}{4}$ **7.** $\frac{15}{2}$ **8.** $\frac{17}{2}$

9. $\frac{45}{4}$ **10.** $\frac{23}{2}$ **11.** $\frac{25}{8}$ **12.** $\frac{27}{8}$

13. $\frac{11}{2}$ **14.** $\frac{23}{4}$ **15.** $\frac{29}{4}$ **16.** $\frac{31}{4}$

17. $\frac{21}{10}$ **18.** $\frac{33}{10}$ **19.** $\frac{41}{10}$ **20.** $\frac{4}{3}$

21. $\frac{10}{3}$ **22.** $\frac{17}{3}$ **23.** $\frac{25}{3}$ **24.** $\frac{32}{3}$

25. $\frac{16}{5}$ **26.** $\frac{28}{3}$ **27.** $\frac{43}{5}$ **28.** $\frac{7}{2}$

29. $\frac{23}{2}$ **30.** $\frac{49}{4}$ **31.** $\frac{29}{10}$ **32.** $\frac{16}{5}$

33. $\frac{57}{8}$ **34.** $\frac{51}{8}$ **35.** $\frac{47}{8}$ **36.** $\frac{29}{10}$

37. $\frac{77}{8}$ **38.** $\frac{37}{10}$ **39.** $\frac{15}{8}$ **40.** $\frac{41}{10}$

Exercise 23

1. $\frac{1}{2}$ **2.** $\frac{4}{3}$ **3.** $\frac{8}{10}$ **4.** $\frac{5}{8}$

5. $\frac{3}{2}$ **6.** 1 **7.** 1 **8.** $\frac{4}{10}$

9. $\frac{5}{4}$ **10.** $\frac{9}{24}$ **11.** $\frac{9}{8}$ **12.** $\frac{6}{5}$

13. $\frac{6}{16}$ **14.** $\frac{5}{16}$ **15.** $\frac{12}{16}$ **16.** $\frac{20}{16}$

17. $\frac{9}{16}$ **18.** $\frac{15}{16}$ **19.** $\frac{11}{10}$ **20.** $\frac{6}{5}$

Exercise 24

1. $\frac{7}{10}$ **2.** $\frac{7}{10}$ **3.** $\frac{11}{10}$ **4.** $\frac{5}{8}$

5. $\frac{5}{8}$ **6.** $\frac{11}{8}$ **7.** $\frac{9}{10}$ **8.** $\frac{10}{15}$

9. $\frac{5}{20}$ **10.** $\frac{13}{8}$ **11.** $\frac{5}{8}$ **12.** $\frac{7}{8}$

13. $\frac{5}{40}$ **14.** $\frac{13}{20}$ **15.** $\frac{11}{8}$ **16.** $\frac{9}{10}$

17. $\frac{13}{8}$ **18.** $\frac{3}{30}$ **19.** $\frac{5}{40}$ **20.** $\frac{8}{40}$

Exercise 25

1. $\frac{1}{2}$
2. $\frac{1}{4}$
3. $\frac{1}{4}$
4. $\frac{1}{5}$

5. $\frac{1}{5}$
6. $\frac{1}{4}$
7. $\frac{1}{4}$
8. $\frac{1}{8}$

9. $\frac{1}{20}$
10. $\frac{3}{16}$
11. $\frac{1}{16}$
12. $\frac{1}{16}$

13. $\frac{3}{20}$
14. $\frac{1}{2}$
15. $\frac{2}{3}$
16. $\frac{1}{12}$

17. $\frac{7}{16}$
18. $\frac{1}{20}$
19. $\frac{1}{20}$
20. $\frac{3}{20}$

Exercise 26

1. $2\frac{3}{4}$
2. $\frac{1}{4}$
3. $5\frac{3}{4}$
4. $\frac{3}{4}$

5. $3\frac{3}{8}$
6. $\frac{7}{8}$
7. $7\frac{5}{8}$
8. $3\frac{3}{8}$

9. $8\frac{1}{4}$
10. $1\frac{1}{4}$
11. $10\frac{3}{4}$
12. $4\frac{1}{4}$

13. $10\frac{3}{4}$
14. $5\frac{3}{4}$
15. $8\frac{3}{8}$
16. $4\frac{1}{8}$

17. $5\frac{3}{4}$
18. $\frac{3}{4}$
19. $6\frac{3}{16}$
20. $2\frac{1}{16}$

Exercise 27

1. $\frac{1}{8}$
2. $\frac{3}{32}$
3. $\frac{2}{25}$
4. $1\frac{1}{4}$

5. 4
6. $\frac{1}{16}$
7. 2
8. 8

9. $3\frac{1}{3}$
10. 4
11. $\frac{1}{24}$
12. $\frac{3}{32}$

13. $\frac{1}{4}$
14. $\frac{1}{2}$
15. $\frac{1}{4}$
16. $\frac{2}{3}$

17. $\frac{4}{9}$
18. $\frac{1}{4}$
19. $\frac{3}{100}$
20. $\frac{2}{5}$

21. $\frac{5}{12}$
22. $\frac{1}{2}$
23. $\frac{1}{12}$
24. $\frac{1}{16}$

25. $\frac{1}{5}$
26. $\frac{3}{7}$
27. $\frac{1}{32}$
28. $\frac{1}{20}$

29. $\frac{3}{20}$
30. $\frac{1}{5}$
31. $\frac{1}{3}$
32. $\frac{1}{9}$

33. 6
34. 9
35. 10
36. $\frac{1}{5}$

37. $12\frac{1}{2}$
38. $\frac{1}{5}$
39. $1\frac{1}{2}$
40. 1

145

Exercise 28

1. 6	**2.** 6	**3.** $6\frac{5}{12}$	**4.** 2
5. 2	**6.** 1	**7.** $1\frac{4}{5}$	**8.** 7
9. $2\frac{5}{8}$	**10.** 10	**11.** 2	**12.** $4\frac{1}{2}$
13. $1\frac{2}{3}$	**14.** $1\frac{4}{5}$	**15.** $2\frac{4}{25}$	**16.** $10\frac{1}{2}$
17. $1\frac{3}{4}$	**18.** $1\frac{1}{32}$	**19.** 32	**20.** 4

Exercise 29

1. $10\frac{2}{3}$	**2.** 128	**3.** 6	**4.** $1\frac{1}{3}$
5. $\frac{3}{4}$	**6.** 2	**7.** 2	**8.** $\frac{1}{2}$
9. $4\frac{1}{2}$	**10.** 8	**11.** 12	**12.** $7\frac{1}{2}$
13. $\frac{1}{8}$	**14.** $1\frac{29}{55}$	**15.** $\frac{2}{3}$	**16.** 4
17. $\frac{3}{4}$	**18.** $1\frac{1}{9}$	**19.** $2\frac{1}{2}$	**20.** $2\frac{1}{8}$

Exercise 30

1. £2	**2.** £9	**3.** 75	**4.** £2
5. 5	**6.** 9	**7.** £35	**8.** £3
9. £6	**10.** £20	**11.** 6 kg	**12.** 12 kg
13. 100	**14.** £350	**15.** 60 kg	**16.** 900 l
17. £9	**18.** £20	**19.** £36	**20.** 250 kg

Exercise 32

1. $\frac{1}{5}$	**2.** $\frac{31}{100}$	**3.** $\frac{101}{500}$	**4.** $\frac{109}{500}$
5. $\frac{1}{10}$	**6.** $\frac{1}{100}$	**7.** $\frac{1}{500}$	**8.** $\frac{7}{20}$
9. $\frac{7}{200}$	**10.** $\frac{41}{100}$	**11.** $\frac{201}{1000}$	**12.** $\frac{23}{50}$
13. $\frac{211}{500}$	**14.** $\frac{1}{2}$	**15.** $\frac{13}{25}$	**16.** $\frac{177}{250}$
17. $\frac{31}{50}$	**18.** $\frac{613}{1000}$	**19.** $\frac{71}{100}$	**20.** $\frac{99}{125}$

Exercise 33

1.	0·3	**2.**	0·47	**3.**	0·023	**4.**	0·0001
5.	0·22	**6.**	0·031	**7.**	0·5	**8.**	0·29
9.	0·6	**10.**	0·03	**11.**	0·003	**12.**	0·05
13.	0·007	**14.**	0·075	**15.**	0·0753	**16.**	0·35
17.	0·0302	**18.**	0·005	**19.**	0·029	**20.**	0·0321

Exercise 34

1.	34·52	**2.**	34·44	**3.**	6·048	**4.**	3·276
5.	80·33	**6.**	7·076	**7.**	57·448	**8.**	1·3
9.	3·888	**10.**	0·574	**11.**	0·58	**12.**	14·538
13.	1·25	**14.**	260·31	**15.**	192·88	**16.**	0·99
17.	0·214	**18.**	11·01	**19.**	2·418	**20.**	0
21.	48·11	**22.**	9·629	**23.**	0·92	**24.**	0·481
25.	11·4	**26.**	1	**27.**	10·5	**28.**	0·22
29.	0·91	**30.**	16·76	**31.**	1·6	**32.**	6·17
33.	9·63	**34.**	0·001	**35.**	93·24	**36.**	39·6
37.	1·01	**38.**	21·86	**39.**	0·001	**40.**	8·1

Exercise 35

1.	24·8	**2.**	0·32	**3.**	248·1	**4.**	0·36
5.	486	**6.**	0·2	**7.**	37·2	**8.**	7
9.	5280	**10.**	39·1	**11.**	21	**12.**	360
13.	82·4	**14.**	1921	**15.**	0·24	**16.**	0·1
17.	9460	**18.**	9·46	**19.**	28 000	**20.**	0·26

Exercise 36

1.	0·278	**2.**	0·2406	**3.**	0·0489	**4.**	0·0692
5.	49·6	**6.**	0·0241	**7.**	0·0334	**8.**	0·00292
9.	0·000436	**10.**	0·407	**11.**	0·0072	**12.**	0·2642
13.	0·84	**14.**	0·0962	**15.**	0·0834	**16.**	0·000279
17.	0·481	**18.**	0·00296	**19.**	0·000058	**20.**	0·001

Exercise 37

1. 4·32	2. 3·84	3. 0·0016	4. 0·06
5. 0·046	6. 0·0036	7. 0·308	8. 3·12
9. 2·25	10. 1·44	11. 1·32	12. 13·34
13. 5·06	14. 11·76	15. 6·05	16. 0·0021
17. 0·0192	18. 0·00102	19. 4·56	20. 0·69
21. 13·13	22. 0·92	23. 17·7	24. 12·4
25. 1·76	26. 6·12	27. 9·18	28. 1·68
29. 8·84	30. 0·09	31. 0·1	32. 0·48
33. 0·048	34. 0·048	35. 0·08	36. 0·32
37. 0·024	38. 0·00001	39. 0·036	40. 0·56

Exercise 38

1. 1·81	2. 7·9	3. 0·0028	4. 0·007
5. 0·4	6. 0·33	7. 0·9	8. 0·64
9. 0·0002	10. 0·344	11. 0·743	12. 29·2
13. 1·11	14. 1·37	15. 0·001	16. 0·33444
17. 1·75	18. 1·05	19. 288·1	20. 1·74
21. 1·1	22. 2·2	23. 5·1	24. 0·7
25. 0·8	26. 1·8	27. 1·06	28. 2·22
29. 1·7	30. 0·0041		

Exercise 39

1. $\frac{1}{100}$ 2. $\frac{1}{50}$ 3. $\frac{1}{25}$ 4. $\frac{3}{50}$ 5. $\frac{2}{25}$ 6. $\frac{1}{10}$ 7. $\frac{3}{25}$

8. $\frac{7}{50}$ 9. $\frac{3}{20}$ 10. $\frac{4}{25}$ 11. $\frac{9}{50}$ 12. $\frac{1}{5}$ 13. $\frac{11}{50}$ 14. $\frac{6}{25}$

15. $\frac{1}{4}$ 16. $\frac{13}{50}$ 17. $\frac{7}{25}$ 18. $\frac{3}{10}$ 19. $\frac{8}{25}$ 20. $\frac{17}{50}$ 21. $\frac{7}{20}$

22. $\frac{9}{25}$ 23. $\frac{19}{50}$ 24. $\frac{2}{5}$ 25. $\frac{21}{50}$ 26. $\frac{11}{25}$ 27. $\frac{9}{20}$ 28. $\frac{23}{50}$

29. $\frac{12}{25}$ 30. $\frac{1}{20}$ 31. $\frac{13}{25}$ 32. $\frac{27}{50}$ 33. $\frac{11}{20}$ 34. $\frac{14}{25}$ 35. $\frac{29}{50}$

36. $\frac{3}{50}$ 37. $\frac{31}{50}$ 38. $\frac{16}{25}$ 39. $\frac{13}{20}$ 40. $\frac{33}{50}$ 41. $\frac{17}{25}$ 42. $\frac{7}{10}$

43. $\frac{18}{25}$ 44. $\frac{37}{50}$ 45. $\frac{3}{4}$ 46. $\frac{19}{25}$ 47. $\frac{39}{50}$ 48. $\frac{4}{5}$ 49. $\frac{41}{50}$

50. $\frac{21}{25}$ 51. $\frac{17}{20}$ 52. $\frac{43}{50}$ 53. $\frac{22}{25}$ 54. $\frac{9}{10}$ 55. $\frac{23}{25}$ 56. $\frac{47}{50}$

57. $\frac{19}{20}$ 58. $\frac{24}{25}$ 59. $\frac{49}{50}$ 60. 1 61. $1\frac{1}{2}$ 62. 2 63. $2\frac{1}{2}$

64.	3	**65.**	$3\frac{1}{2}$	**66.**	4	**67.**	$4\frac{1}{2}$	**68.**	$1\frac{1}{10}$	**69.**	$1\frac{1}{5}$	**70.**	$1\frac{3}{5}$
71.	$2\frac{1}{10}$	**72.**	$2\frac{1}{5}$	**73.**	$2\frac{4}{5}$	**74.**	$3\frac{2}{5}$	**75.**	$3\frac{1}{10}$	**76.**	$3\frac{1}{5}$	**77.**	5
78.	$5\frac{1}{2}$	**79.**	$2\frac{9}{10}$	**80.**	$2\frac{3}{4}$								

Exercise 40

1.	£40	**2.**	2 tonnes	**3.**	£5	**4.**	30	**5.**	£4
6.	£77	**7.**	9 cm	**8.**	2 cm	**9.**	50 cm^3	**10.**	£3
11.	£200	**12.**	18 m	**13.**	50 000	**14.**	£30	**15.**	1 km
16.	2 km	**17.**	6 kg	**18.**	£700	**19.**	£400	**20.**	£6·50

Exercise 41

1.	Loss £1	**2.**	Gain £1	**3.**	Gain 50p	**4.**	Gain £1
5.	Gain 50p	**6.**	Gain £2	**7.**	Gain 25p	**8.**	Loss £1·20
9.	Gain 10p	**10.**	Loss £2·50	**11.**	Gain £5·50	**12.**	Gain £1·20
13.	Gain 20p	**14.**	Loss £1·50	**15.**	Loss £3·05		

Exercise 42

1.	Gain £2·10	**2.**	Gain 20p	**3.**	Loss £3·28	**4.**	Gain £1
5.	Loss £3·60	**6.**	Loss £4·80	**7.**	Gain £4·80	**8.**	Loss £16
9.	Loss £60	**10.**	Gain £2·00	**11.**	Gain £60	**12.**	Loss £1·20
13.	Loss £3	**14.**	Gain £56	**15.**	Gain 6p		

Exercise 43

1.	£20	**2.**	£90	**3.**	£200	**4.**	£120	**5.**	£360	**6.**	£90
7.	£540	**8.**	£400	**9.**	£125	**10.**	£160	**11.**	£600	**12.**	£960
13.	£1000	**14.**	£80	**15.**	£400						

Exercise 44

1. 84 marks
2. £16·92
3. £3
4. £2000
5. 200 rupees
6. £4
7. 770 pes
8. 3680 fr
9. £60
10. £24
11. 223 rupees
12. £17
13. £200
14. £40
15. 204 dollars

Exercise 45

1. £9·40
2. £25·20
3. £20
4. £190
5. £24·40
6. £90
7. £74
8. £68·80
9. £750
10. £430
11. £12·80
12. £36

Exercise 46

1. £118
2. £154·50
3. £214
4. £127·50
5. £88
6. £188
7. £122
8. £174
9. £204
10. £315
11. £153
12. £180
13. £119
14. £169·50
15. £196

Exercise 47

1. 80 km
2. 120 km
3. 20 km
4. 10 km
5. 240 km
6. 400 km
7. 140 km
8. $93\frac{1}{3}$ km
9. 180 km
10. 350 km

Exercise 48

1. 5 hrs
2. 10 hrs
3. 4 hrs
4. 20 hrs
5. 2 hrs
6. 3 hrs 20 mins
7. $2\frac{1}{2}$ hrs
8. 8 hrs
9. 2 hrs 40 mins
10. 6 hrs 40 mins

Exercise 49

1. 4 hrs
2. 4 hrs
3. 80 km
4. 50 km/hr
5. 3 hrs
6. 12 km/hr
7. 400 km
8. 10 hrs
9. 1000 km/hr
10. 10 hrs
11. 320 km
12. 20 km/hr
13. 10 hrs
14. 4 hrs
15. 10 km/hr
16. 36 km

Exercise 50

1. 200 cm	**2.** 350 cm	**3.** 1000 cm	**4.** 500 cm
5. 240 cm	**6.** 2000 m	**7.** 10 000 m	**8.** 15 000 m
9. 20 000 m	**10.** 22 000 m	**11.** 2 m	**12.** 4 m
13. 40 m	**14.** 80 m	**15.** 800 m	**16.** 9 m
17. 2 km	**18.** 10 km	**19.** 5 km	**20.** 3·5 km

Exercise 51

1. 30 cm^2	**2.** 12 cm^2	**3.** 6 cm	**4.** 32 m^2	**5.** 5 m
6. 1·6 m	**7.** 8 m	**8.** 4·5 cm^2	**9.** 180 cm^2	**10.** 8 m
11. 12 m^2	**12.** 12 cm	**13.** 5 cm	**14.** 2 m^2	**15.** 7 m^2
16. 3 m	**17.** 33 cm^2	**18.** 3 cm	**19.** 34 m^2	**20.** 21 m^2

Exercise 52

1. 6 ha	**2.** 8 ha	**3.** 5 ha	**4.** 150 ha	**5.** 56 ha
6. 13·5 ha	**7.** 13 ha	**8.** 1·28 ha	**9.** 7·6 ha	**10.** 80 ha

Exercise 53

1. 6 cm^3	**2.** 192 cm^3	**3.** 2 million cm^3	**4.** 1·5 m^3
5. 3000 cm^3	**6.** 20 000 cm^3	**7.** $\frac{1}{64}$ m^3	**8.** 6 m^3
9. 100 m^3	**10.** 600 m^3	**11.** 1 m^3	**12.** 1000 cm^3
13. 200 cm^3	**14.** 8 m^3	**15.** 400 m^3	

Exercise 54

1. 24 litres	**2.** 100 litres	**3.** 32 litres	**4.** 720 litres
5. 384 litres	**6.** 45 litres	**7.** 1 litre	**8.** 120 litres
9. 9 litres	**10.** 114 litres	**11.** 5 litres	**12.** 0·6 litres

Exercise 55

1. 2000 g	**2.** 2000 kg	**3.** 5·94 kg	**4.** 0·826 tonnes

5.	0·426 kg	6.	4500 kg	7.	3200 kg	8.	2 kg
9.	3·5 tonnes	10.	8000 g	11.	2480 g	12.	3800 kg
13.	1200 g	14.	10 200 kg	15.	32·469 kg	16.	8400 g
17.	5000 kg	18.	1 200 000 g	19.	500 000 g	20.	18 600 g

Answers to Algebra Exercises

Exercise 1

1. 3	2. 5	3. 4	4. 11	5. 20
6. 18	7. 38	8. 18	9. 22	10. 26
11. 30	12. 23	13. 32	14. 19	15. 28
16. 28	17. 17	18. 29	19. 50	20. 54
21. 111	22. 46	23. $9\frac{1}{2}$	24. 106	25. 59
26. $25\frac{1}{2}$	27. 116	28. 40	29. 61	30. 7
31. 110	32. 102	33. 1	34. 1	35. 99
36. 604	37. 133	38. $\frac{4}{5}$	39. $4\frac{1}{4}$	40. 1100

Exercise 2

1. 4	2. 3	3. 1	4. 1	5. 18
6. 15	7. 24	8. 5	9. 20	10. 25
11. 10	12. 31	13. 22	14. 9	15. 23
16. 22	17. 26	18. 4	19. 4	20. 2
21. $2\frac{1}{2}$	22. $1\frac{3}{4}$	23. $\frac{1}{2}$	24. 76	25. 900
26. 6	27. $3\frac{1}{2}$	28. $2\frac{1}{2}$	29. $2\frac{1}{2}$	30. $8\frac{3}{4}$
31. 73	32. $\frac{1}{2}$	33. $\frac{3}{4}$	34. 1	35. 37
36. 9	37. 7	38. 6	39. 4	40. 6

Exercise 3

1. 8	2. 12	3. 24	4. 20	5. 28
6. 2	7. 1	8. 32	9. 3	10. 40
11. 36	12. 52	13. 44	14. 48	15. $\frac{1}{2}$
16. 80	17. $\frac{4}{5}$	18. $\frac{2}{5}$	19. 120	20. 400
21. 10	22. 7	23. 12	24. 8	25. 15
26. 3	27. 4	28. 2	29. 1	30. 22
31. 31	32. 19	33. 20	34. 30	35. 2
36. 53	37. 0	38. 20	39. 4	40. 0

Exercise 4

1. 7	**2.** 11	**3.** 1	**4.** 13	**5.** 21
6. 5	**7.** 8	**8.** 11	**9.** 11	**10.** 9
11. 21	**12.** 23	**13.** 17	**14.** 13	**15.** 9
16. 22	**17.** 10	**18.** 27	**19.** 33	**20.** 18
21. 7	**22.** 19	**23.** 4	**24.** 44	**25.** 1
26. 0	**27.** 0	**28.** 0	**29.** 3	**30.** 55
31. 30	**32.** 0	**33.** 1	**34.** 31	**35.** 1
36. $\frac{1}{2}$	**37.** 12	**38.** 49	**39.** 1	**40.** 1

Exercise 5

1. 12	**2.** 12	**3.** 18	**4.** 18	**5.** 24
6. 3	**7.** 2	**8.** 48	**9.** 54	**10.** 9
11. 27	**12.** 33	**13.** 60	**14.** 120	**15.** 1
16. 30	**17.** 15	**18.** 240	**19.** 600	**20.** 300

Exercise 6

1. 7	**2.** 11	**3.** 14	**4.** 8	**5.** 22
6. 1	**7.** 26	**8.** 14	**9.** 26	**10.** 7
11. 34	**12.** 25	**13.** 45	**14.** 22	**15.** 7
16. 23	**17.** 38	**18.** 70	**19.** 38	**20.** 46

Exercise 7

1. 5	**2.** 3	**3.** 1	**4.** 1	**5.** 10
6. 1	**7.** 2	**8.** 2	**9.** 2	**10.** 4
11. 7	**12.** 5	**13.** 2	**14.** 3	**15.** 2
16. 12	**17.** 4	**18.** 4	**19.** 1	**20.** 2

Exercise 8

1. 6	**2.** 2·0	**3.** 7	**4.** 1	**5.** 8
6. 2	**7.** 10	**8.** 6	**9.** 10	**10.** 17

| 11. 8 | 12. 19 | 13. 31 | 14. 11 | 15. 5 |
| 16. 29 | 17. 9 | 18. 18 | 19. 8 | 20. 18 |

Exercise 9

1. 16	2. 11	3. 4	4. 48	5. 48
6. 12	7. 26	8. 17	9. 14	10. 1
11. 6	12. 5	13. 40	14. 24	15. 17
16. 20	17. 24	18. 10	19. 33	20. 17

Exercise 10

1. 1	2. 2	3. 3	4. 1	5. 18
6. $1\frac{1}{2}$	7. $1\frac{1}{3}$	8. 2	9. 2	10. 2
11. 48	12. 12	13. $\frac{1}{5}$	14. 24	15. 11
16. 8	17. 26	18. 30	19. 30	20. 20

Exercise 11

1. 22	2. 2	3. 2	4. 3	5. 1
6. 8	7. 17	8. 23	9. 1	10. 3
11. 16	12. 27	13. $\frac{1}{2}$	14. 15	15. 16
16. 25	17. 42 .	18. 1	19. 55	20. $\frac{1}{2}$

Exercise 12

1. $\dfrac{p + t}{x}$ 2. $q - p$ 3. $c - 2m$

4. $\dfrac{b - a}{p}$ 5. $a + b + t$ 6. $2t + 4q$

7. $v(t + w)$ 8. $2t - m$ 9. $cd - 5$

10. $\dfrac{p}{q}$ 11. $\dfrac{p + q}{t}$ 12. $\dfrac{x + y}{x - y}$

13. $4d - 2q$ 14. $3m + 5t$ 15. $\dfrac{2r}{3p}$

155

16. $t(x + r)$ **17.** $tq + 3$ **18.** $\dfrac{p}{q} - 5$

19. $m(a + b + c)$ **20.** $(a + b) - x$

Exercise 13

1. $4x$	**2.** $14y$	**3.** $8t$	**4.** $10x$
5. $16a$	**6.** $16x$	**7.** $13c$	**8.** $11x$
9. $14y$	**10.** $17x$	**11.** $35x$	**12.** $15p$
13. $15a$	**14.** $20x$	**15.** $27a$	**16.** $17x$
17. $25x$	**18.** $10y$	**19.** $16t$	**20.** $14t$

Exercise 14

1. $x + y$	**2.** $21x$	**3.** $a + y$	**4.** $a + 2y$
5. $ab + c$	**6.** $3a + 3c$	**7.** $11x + 2y$	**8.** $x + 3y$
9. $15x + 3a$	**10.** $9a + 3b$	**11.** $8a + 8b$	**12.** $13x + y$
13. $3a + 3b$	**14.** $5x + 14y$	**15.** $3x + 20y$	**16.** $3a + 8b$
17. $26x - y$	**18.** $x + y$	**19.** $6x + y$	**20.** $12a + 8c$

Exercise 15

1. 4	**2.** 8	**3.** 6	**4.** 12	**5.** 4
6. 16	**7.** 8	**8.** 32	**9.** 12	**10.** 48
11. 2	**12.** 8	**13.** 1	**14.** 20	**15.** 80
16. 16	**17.** 64	**18.** 64	**19.** 128	**20.** 192

Exercise 16

1. 4	**2.** 16	**3.** 36	**4.** 8	**5.** 64
6. 16	**7.** 96	**8.** 108	**9.** 96	**10.** 48
11. 144	**12.** 72	**13.** 16	**14.** 48	**15.** 64
16. 48	**17.** 48	**18.** 44	**19.** 70	**20.** 100

Exercise 17

1. 11	**2.** 22	**3.** 16	**4.** 4	**5.** $6\frac{1}{6}$
6. 3	**7.** 54	**8.** 72	**9.** 27	**10.** $36\frac{1}{2}$
11. 6	**12.** 2	**13.** 7	**14.** $\frac{1}{2}$	**15.** $\frac{1}{2}$
16. $12\frac{2}{3}$	**17.** 5	**18.** $1\frac{1}{6}$	**19.** 8	**20.** 81

Exercise 18

1. 2	**2.** 16	**3.** 6	**4.** 6	**5.** 10
6. 2	**7.** 1	**8.** 1	**9.** $\frac{3}{4}$	**10.** 48
11. 16	**12.** 20	**13.** 96	**14.** 20	**15.** 46
16. 64	**17.** 1	**18.** 16	**19.** 4	**20.** 8

Exercise 22

1. $6px$	**2.** $6px$	**3.** $32ab$	**4.** $6ab$
5. $10ap$	**6.** $6ap$	**7.** $8b^2$	**8.** $10y^2$
9. $12m^2$	**10.** $64t^2$	**11.** $60p^2$	**12.** $48a^2$
13. $10abc$	**14.** $8abd$	**15.** $10abc$	**16.** $6txy$
17. $14ab$	**18.** $60pqr$	**19.** $60p^2$	**20.** $84x^2$
21. $15am$	**22.** $35p^2$	**23.** $20t^2$	**24.** $80abc$
25. $10cd$	**26.** $21ct$	**27.** $36apt$	**28.** $30cmt$
29. $16ab$	**30.** $6mc$	**31.** $32cd$	**32.** $16cd$
33. $10mr$	**34.** $12am$	**35.** $21t^2$	**36.** $40ce$
37. $36mr$	**38.** $48t^2$	**39.** $10ap$	**40.** $8abc$

Exercise 23(a)

1. 4	**2.** 3	**3.** 15	**4.** 4
5. 9	**6.** 42	**7.** 4	**8.** 1
9. 9	**10.** 12	**11.** 3	**12.** 9
13. 12	**14.** 4	**15.** 18	**16.** 12
17. 16	**18.** 36	**19.** 144	**20.** 60
21. 110	**22.** 72	**23.** 36	**24.** 12
25. 20	**26.** 54	**27.** 36	**28.** 144

29. 4	**30.** 8	**31.** 72	**32.** 27
33. 54	**34.** 22	**35.** 3	**36.** 1
37. 4	**38.** 60	**39.** 3	**40.** 216

Exercise 23(b)

1. 2	**2.** 9	**3.** 10	**4.** 6
5. 18	**6.** 14	**7.** 1	**8.** 9
9. 4	**10.** 12	**11.** 18	**12.** 12
13. 9	**14.** 18	**15.** 12	**16.** 6
17. 24	**18.** 12	**19.** 144	**20.** 60
21. 165	**22.** 72	**23.** 36	**24.** 54
25. 15	**26.** 12	**27.** 216	**28.** 16
29. 9	**30.** 9	**31.** 4	**32.** 72
33. 48	**34.** 33	**35.** 54	**36.** 1
37. $\frac{1}{2}$	**38.** 60	**39.** 81	**40.** 12

Exercise 24

1. $5a$	**2.** $3y$	**3.** $2xy$	**4.** $2m$
5. $3p$	**6.** $5a$	**7.** $2pq$	**8.** $ab/2$
9. $5m$	**10.** $2a$	**11.** $\frac{1}{2}a$	**12.** $5q$
13. $5a/x$	**14.** $3a/b$	**15.** $5m/a$	**16.** 1
17. $2a^2/b$	**18.** $3pq$	**19.** $2a$	**20.** $2a^2$
21. $2ab$	**22.** $2c^2$	**23.** $5de$	**24.** x/y
25. $3p^2$	**26.** $2a^2b$	**27.** $x^2/2$	**28.** $5a$
29. $2am$	**30.** $\dfrac{p^3}{2}$	**31.** $d/2$	**32.** $6e/d$
33. $2t$	**34.** $36t/a$	**35.** $ty/2$	**36.** $2p$
37. $3m$	**38.** $2p^2$	**39.** $5a$	**40.** $2x/t$

Exercise 25

1. 12 cm; 8 cm²	**2.** 20; 24	**3.** 4; 1	**4.** 14; 12
5. 34; 70	**6.** 38; 90	**7.** 56; 160	**8.** 60; 200
9. 54; 182	**10.** 42; 98	**11.** 17; 15	**12.** 17; $17\frac{1}{2}$

13. 7; 3	**14.** 25; $38\frac{1}{2}$	**15.** 100; 600	**16.** 88; 480
17. 56; 180	**18.** 63; $247\frac{1}{2}$	**19.** 48; 140	**20.** 52; 168
21. 29; 52	**22.** 31; 60	**23.** 21; $13\frac{1}{2}$	**24.** 30; 54
25. 27; $40\frac{1}{2}$	**26.** 35; 75	**27.** 25; 25	**28.** 70; 300
29. 27; 35	**30.** 6; 2	**31.** 64; 255	**32.** 53; 165
33. 15; 14	**34.** 57; 200	**35.** 84; 432	**36.** 104; 640
37. 96; 512	**38.** 16; $13\frac{3}{4}$	**39.** 17; 14	**40.** 21; $22\frac{1}{2}$

Exercise 26

1. 30 km	**2.** 120	**3.** 50	**4.** 240
5. 80	**6.** 100	**7.** 80	**8.** 160
9. 200	**10.** 300	**1.** 140	**12.** 280
13. 125	**14.** 140	**15.** 800	**16.** 680
17. 180	**18.** 300	**19.** 30	**20.** 25
21. 360	**22.** 27	**23.** 70	**24.** 255
25. 220	**26.** 320	**27.** 54	**28.** 120
29. 225	**30.** 288	**31.** 496	**32.** 125
33. 23	**34.** 25	**35.** 21	**36.** 192
37. 13	**38.** 9	**39.** 112	**40.** 136

Exercise 27

1. 5	**2.** 7	**3.** 10	**4.** 9
5. 21	**6.** 10	**7.** 14	**8.** 20
9. 86	**10.** 15	**11.** 66	**12.** 26
13. 15	**14.** 11	**15.** 32	**16.** 41
17. 232	**18.** 38	**19.** 36	**20.** 36
21. 9	**22.** 17	**23.** 32	**24.** 19
25. 32	**26.** 201	**27.** 84	**28.** 142
29. 73	**30.** 26	**31.** 27	**32.** 32
33. 34	**34.** 6	**35.** 9	**36.** 2006
37. 132	**38.** 1001	**39.** 29	**40.** 1002

Exercise 28

1. $\frac{2}{15}$	**2.** 1	**3.** 1	**4.** $\frac{9}{17}$
5. 1	**6.** 2	**7.** $1\frac{2}{3}$	**8.** $\frac{9}{14}$
9. $1\frac{1}{24}$	**10.** $2\frac{2}{9}$	**11.** $\frac{2}{3}$	**12.** $\frac{2}{3}$
13. $1\frac{1}{7}$	**14.** $\frac{4}{5}$	**15.** $\frac{1}{7}$	**16.** 2
17. 2	**18.** $3\frac{1}{3}$	**19.** $2\frac{11}{12}$	**20.** 5
21. $1\frac{3}{7}$	**22.** $\frac{1}{12}$	**23.** $\frac{1}{3}$	**24.** $1\frac{1}{4}$
25. 8	**26.** 2	**27.** 2	**28.** 5
29. 3	**30.** 2	**31.** 2	**32.** 10
33. $1\frac{1}{2}$	**34.** 3	**35.** 20	**36.** 0
37. 1	**38.** 3	**39.** 2	**40.** 1

Exercise 29

1. $6a + 3c$	**2.** $2b + c$	**3.** $30a + 45c$	**4.** $7x + 17y$
5. $24x - 8y$	**6.** $12a + 30c$	**7.** $8x + 32y$	**8.** $10p + 5q$
9. $4a + 36d$	**10.** $18x - 6y$	**11.** $20x + 4y$	**12.** $9x + 60y$
13. $4x - 24y$	**14.** $40x - 4y$	**15.** $4x + 10a$	**16.** $8p + 56q$
17. $12p - 30q$	**18.** $30x + 10y$	**19.** $40x - 30y$	**20.** $35x - 30y$
21. $10p + 5q$	**22.** $7p - 7q$	**23.** $20p + 10$	**24.** $15p - 5q$
25. $6a + 18b$	**26.** $14a - 7b$	**27.** $12a - 18b$	**28.** $14a + 7b$
29. $6a - 18b$	**30.** $3x + 6y$	**31.** $10y - 5t$	**32.** $6a + 18t$
33. $4a - 2b$	**34.** $15m + 10n$	**35.** $6m - 18n$	**36.** $16a + 88b$
37. $8a - 6c$	**38.** $3a + 4c$	**39.** $b + 2t$	**40.** $20b + 30a$
41. $10m + 15t$	**42.** $18a - 3b$	**43.** $10m + 22$	**44.** $6t + 12a$
45. $6 + 15a$	**46.** $28a - 4c$	**47.** $10a + 12c$	**48.** $15a - 3r$
49. $30a + 5$	**50.** $14a - 6$	**51.** $18t + 15$	**52.** $4x + 32y$
53. $6x - 18y$	**54.** $4x - 72$	**55.** $3x + 4y$	**56.** $30a - 10d$
57. $36a + 12d$	**58.** $10 + 20a$	**59.** $10 + 40t$	**60.** $2m + 20t$

Exercise 30

1. 1	**2.** 1	**3.** 3	**4.** 2
5. 5	**6.** 17	**7.** 3	**8.** 1
9. 13	**10.** 12	**11.** 8	**12.** 10
13. 27	**14.** 19	**15.** 7	**16.** 17
17. 12	**18.** 4	**19.** 21	**20.** 45
21. 6	**22.** 10	**23.** 31	**24.** 33
25. 93	**26.** 83	**27.** 34	**28.** 72
29. 105	**30.** 101	**31.** 20	**32.** 6
33. 26	**34.** 19	**35.** 9	**36.** 20
37. 41	**38.** 20	**39.** 93	**40.** 30

Exercise 31

1. 4	**2.** 7	**3.** 11	**4.** 23
5. 14	**6.** 22	**7.** 15	**8.** 19
9. 12	**10.** 22	**11.** 10	**12.** 101
13. 17	**14.** 21	**15.** 42	**16.** 19
17. 28	**18.** 6	**19.** 61	**20.** 57
21. 18	**22.** 17	**23.** 1	**24.** 8
25. 6	**26.** 11	**27.** 9	**28.** 17
29. 1	**30.** 8	**31.** 21	**32.** 20
33. 3	**34.** 2	**35.** 10	**36.** 41
37. 26	**38.** 38	**39.** 94	**40.** 70

Exercise 32

1. 23	**2.** 73	**3.** 6	**4.** 113
5. 11	**6.** 15	**7.** 22	**8.** 17
9. 74	**10.** 20	**11.** 105	**12.** 73
13. 24	**14.** 299	**15.** 111	**16.** 41
17. 54	**18.** 113	**19.** 1031	**20.** 78
21. 1	**22.** 16	**23.** 102	**24.** 14
25. 68	**26.** 17	**27.** 52	**28.** 9

29. 39	30. 12	31. 92	32. 35
33. 32	34. 35	35. 99	36. 23
37. 30	38. 30	39. 17	40. 7

Exercise 33

1. 2	2. 3	3. 3	4. 1
5. 3	6. 2	7. 2	8. 4
9. 4	10. 1	11. 5	12. 1
13. 3	14. 1	15. 2	16. 7
17. 10	18. 1	19. 1	20. 0
21. 1	22. 0	23. $\frac{3}{2}$	24. 3
25. 2	26. 2	27. 2	28. 0
29. 13	30. 8	31. $\frac{5}{2}$	32. 4
33. 1	34. 2	35. 34	36. 5
37. 2	38. $\frac{13}{2}$	39. $\frac{7}{2}$	40. $\frac{9}{2}$

Exercise 34

1. 20	2. 1	3. 2	4. 1
5. 1	6. 10	7. 1	8. 3
9. 1	10. 2	11. 1	12. 4
13. 5	14. 5	15. 2	16. 3
17. 1	18. 1	19. 4	20. 3
21. 2	22. 1	23. 2	24. 8
25. 4	26. 12	27. 1	28. 8
29. 2	30. 5		

Answers to Geometry Exercises

Exercise 11

1.	157°	2.	59°	3.	90°	4.	101°	5.	117°
6.	42°	7.	1°	8.	141°	9.	133°	10.	122°
11.	111°	12.	108°	13.	92°	14.	87°	15.	79°
16.	61°	17.	52°	18.	49°	19.	38°	20.	12°

Exercise 12

1.	88°	2.	8°	3.	76°	4.	19°	5.	61°
6.	27°	7.	59°	8.	32°	9.	43°	10.	82°
11.	7°	12.	73°	13.	12°	14.	63°	15.	28°
16.	57°	17.	39°	18.	48°	19.	89°	20.	45°

Exercise 18

1. BEF, 59°
2. GKB = LMF = AKL = CLM = EMN = 114°
   All the other angles in the figure equal 66°.

Exercise 19

1. All the acute angles are 78°. All the obtuse ones are 102°.
4. The acute angles are all 58° and the obtuse ones are all 93°.
5. All the acute angles are 86° and all the obtuse angles are 132°.

Exercise 21

1.	75°	2.	70°	3.	65°
4.	55°	5.	71°	6.	67°
7.	74°	8.	35°	9.	45°

Exercise 22

1. 60°	2. 32°	3. 30°
4. 36°	5. 31°	6. 47°
7. 54°	8. 34°	9. 40°
10. 20°		

Exercise 24

1. 40°	2. 127°; QR	3. 63°; 117°
4. 44°; 68°	5. 53°; 7·2 cm	6. 60°
7. 40°; $12\frac{1}{4}$ cm	8. the largest angle, $68\frac{1}{2}$°	9. 120°
10. 44°; 22°		

Exercise 25

1. 32°	2. 42°	3. 38°
4. 48°	5. 59°	6. 67°
7. 45°	8. 70°	9. 57°
10. 70°		

Exercise 28

1. DC = 5 cm; BC = 3 cm; DB = 6·4 cm; AC = 6·4 cm; ∠BDC = ∠BAC = ∠ACD = 28°; ∠CAD = ∠ADB = ∠DBC = ∠ACB = 62°; ∠AED = 56°; ∠DEC = ∠AEB = 12·4°; A = 15 cm² P = 16 cm

2. PQ = QR = PS = 10 cm;   TP = TQ = TR = 7 cm;   PR = QS = 14 cm; ∠PTQ = ∠QTR = ∠RTS = ∠STP = 90°; ∠QSR = ∠QSP = 45° and so on

3. (a) A = 35 cm²; P = 24 cm   (b) A = 80 cm²; P = 36 cm   (c) A = 121 cm²; P = 44 cm (d) A = 64 cm²; P = 32 cm

4. P = 16·4 cm; A = 15·6 cm²

5. P = 44·8 cm; A = 125 cm²; equal

6. P = 23·2 cm; A = 32·4 cm²; 19·3 cm; the same

7. 5·65 cm

8. No

9. 7 cm approx.; 90°

10. 5 cm; 30 cm; 11·2 cm; 26·2 cm; 25 cm²

Answers to Miscellaneous Questions

1. £150
2. 0·29%
3. $2\frac{17}{32}$
4. £80
5. $\frac{5}{14}$
6. 23
7. £4·40
8. 28 400 m
9. £8·80
10. 13 125 litres
11. 44
12. 10
13. $\frac{100t}{x}$ eggs
14. 13a
15. 2
16. $\frac{100m}{q}$ pence
17. $8p^2q^2$
18. $\frac{ab}{2}$
19. $3\frac{1}{2}$
20. £$\frac{cx}{100}$
21. 157°
22. $15\frac{1}{2}°$
23. right
24. 180°
25. acute, acute, obtuse, reflex, acute
26. 57°
27. 40°, 100°
28. 60°
31. 21p
32. $4\frac{1}{4}$
33. 15%
34. 2 000 004
35. One thousand and one
36. 24·18
37. £1500
38. 4808
39. 160p
40. $\frac{3}{10}$
41. $14xy + x$
42. $24x^3y$
43. $\frac{ad}{2}$
44. 27
45. 18
46. $\frac{x}{t}$ km/h
47. td pence
48. $qt + 3$
49. $\frac{2}{3}$
50. 1
51. 42°
52. 59°
53. 138°
54. AC
55. 61°
56. $3t$ m : $\frac{t^2}{2\,m^2}$
57. (a) 5 cm (b) 66°
58. $(180 - p)°$
59. $(90 - t)°$
60. They are *not* parallel
61. 40%; $12\frac{1}{2}$%; 80%
62. 6
63. 5
64. £141·90
65. 3·945
66. $3\frac{7}{8}$
67. 27%
68. $\frac{1}{2}$
69. 208 tiles
70. $\frac{1}{2}$, $\frac{11}{20}$, $\frac{3}{5}$, $\frac{7}{10}$, $\frac{3}{4}$
71. 58
72. 3
73. 3a
74. $20a^3p^3$
75. $3xy$
76. $2a = c$
77. 10x
78. $3xy$
79. 98
80. 9p
81. $(180 - 2p)°$
82. $(90 - 5x)°$
83. ∠EBC
84. $(180 - x)°$
85. (a) 3 cm (b) 5 cm (c) 102° (d) 78° (e) 102°

86. 220°	**87.** 30°	
88. MT, TR, MR, NT, NP, NQ, TP, TQ, PQ		**89.** 8 cm
90. $x = 18°; 90°, 54°, 36°$		**91.** $\frac{7}{9}$
92. £10 000	**93.** £37·50	**94.** 10
95. 9·36	**96.** 250 000	**97.** 33
98. 24 200 000	**99.** 0·5; 0·25; 0·125; 0·1	**100.** 29
101. 100	**102.** $2mc$	**103.** $64p^2q$
104. $200t$ cm	**105.** x/100 000	**106.** $23x$
107. 8	**108.** 15	**109.** 28
110. $\dfrac{p^2}{2d}$	**111.** $87\frac{1}{2}°$	**112.** obtuse

113. They are *not* parallel	**114.** 40°	**115.** BD
116. 38°	**117.** $x = 100°; a = 48°$	**118.** $x = 106°; a = c = 74°$
119. BC = 3 cm	**120.** 117° or 63°	**121.** $\frac{7}{8}$, $\frac{3}{4}$, $\frac{1}{2}$, $\frac{1}{3}$, $\frac{5}{16}$
122. £50	**123.** 36, 12 years	**124.** $666\frac{2}{3}$ cm^3
125. 3	**126.** $\frac{1}{3}$	**127.** £4·40
128. 4	**129.** 3·374	**130.** 6
131. 17	**132.** 12	**133.** $5t + 8$
134. $9a$	**135.** 26	**136.** $6pq^2$
137. $4a$	**138.** 120	**139.** $19ab + b$
140. 40	**141.** 360°	**142.** $(180 - t)°$
143. 58°	**144.** No	**145.** 102°
146. $(160 - t)°$	**147.** 10 cm	**148.** unequal
149. $x = 45°; y = p = 135°$	**150.** 13	**151.** 21 006
152. two million and four	**153.** 72	**154.** 0·921
155. 1022·1	**156.** 32 m^2	**157.** 12 m^3
158. 20%	**159.** $\frac{1}{2}$	**160.** 50%
161. True	**162.** 10	**163.** 13
164. 13	**165.** 166	**166.** 1 000 000t cm^3
167. £ $\dfrac{cd}{50}$	**168.** $6p$	**169.** $4a$
170. 200	**171.** $(180 - 2p)°$	**172.** 28°

173. 4	**174.** 107°	**175.** acute
176. 360°	**177.** reflex	**178.** $\triangle ABC = 56°$
179. PQ − 3·6 cm	**180.** $(90 - m)°$	**181.** 1
182. 375 cm³	**183.** 102	**184.** 0·042
185. 15	**186.** 5	**187.** 25%
188. $\frac{1}{3}$	**189.** 366 days	**190.** No
191. 11	**192.** 125	**193.** 1
194. 56	**195.** 2	**196.** 8
197. 15	**198.** $2\frac{1}{4}$	**199.** 36 m³
200. $\frac{pm}{2}$	**201.** 270°	**202.** 180°
203. 54°; 35°	**204.** 74°	**205.** triangle law
206. 41°	**207.** 120°	**208.** obtuse
209. 12 cm	**210.** 35°	**211.** £36
212. £1·97	**213.** $\frac{1}{4}$	**214.** $12\frac{1}{2}\%$
215. $\frac{1}{32}$	**216.** 8%	**217.** $\frac{1}{2}$
218. 2·764	**219.** $33\frac{1}{3}\%$	**220.** 29 km/h
221. 6	**222.** 25	**223.** $28a^3b$
224. $4\,m^2n^3$	**225.** $6p$	**226.** $7d + 5$
227. $50x$ pence	**228.** $2\,000\,000\,c$ mg	**229.** 55 km/h
230. $11x$	**231.** 30°	**232.** 76°
233. $(90 - 12t)°$	**234.** $x + y = 180$	**235.** No
236. TN $\doteqdot$ 5·4 cm	**238.** $p + 2t = 180°$	**239.** acute
240. $(x + t)°$	**241.** 60·36	**242.** $\frac{4}{5},\quad \frac{3}{5},\quad \frac{2}{25},\quad \frac{1}{500}$
243. $\frac{2}{5}, \frac{1}{100}$	**244.** 0·7, 0·06, 0·003, 0·64, 0·235	
245. $3\frac{1}{5}$	**246.** $3\frac{1}{3}$	**247.** $\frac{27}{8}$
248. 23 364	**249.** 5·1 cm and 16·1	**250.** 2 m²
251. $1\frac{1}{2}$	**252.** $6a$	**253.** 80
254. $8c^3$	**255.** $2t$	**256.** 4
257. $\frac{t}{60}$ hours	**258.** $7x$	**259.** 25
260. 2	**261.** 40°)	**262.** 49°
263. $(7 - x)$ cm	**264.** 67°	**265.** $3\frac{1}{2}$ cm²

266. $(90 - \frac{1}{2}t)°$

267. $\frac{1}{9}$

268. $\frac{1}{4}$

269. $(x - y)$ cm

270. reflex

271. £4·75

272. $7\frac{1}{5}$

273. 137·65

274. $2\frac{1}{2}\%$

275. 18

276. 7·32

277. 102

278. 0·231

279. £5

280. £1000

281. $34x + 7$

282. $17acpq$

283. $2x$

284. 13

285. $24m$ hrs

286. $\dfrac{t}{1\,000\,000}$ kg

287. $5x$

288. 10

289. 6

290. px km

291. $AC \doteqdot 4·8$ cm

292. $(90 - 3x)°$

293. $\frac{1}{9}$

294. $10°$

296. No

297. $a + x + t = 180°$

298. They are *not* parallel

299. $x = 46, a = c = 134°$

300. $\angle PQR \doteqdot 67°$

168